# How to Be Your Dog's Best Friend

## The Monks of New Skete

# How to Be Your Dog's Best Friend

## THE CLASSIC TRAINING MANUAL FOR DOG OWNERS

*Completely Revised and Updated*

LITTLE, BROWN AND COMPANY
*New York  Boston*

Little, Brown and Company
Time Warner Book Group
1271 Avenue of the Americas, New York, NY 10020
Visit our Web site at www.twbookmark.com

Second Edition

Some material in the original edition of this book appeared
in *Off Lead,* the *National Dog Training Monthly,* and in *Gleanings,*
Journal of New Skete, in slightly different form.

ISBN 0-316-61000-3
LCCN 2002102894

10  9  8  7  6  5  4

Q-FF

Printed in the United States of America

*To our many clients over
the years, whose love for
their dogs has been a constant
source of inspiration for us*

# Also by the Monks of New Skete

DOG CARE AND TRAINING
*The Art of Raising a Puppy*

SPIRITUALITY
*In the Spirit of Happiness*

*Learning the value of silence is learning to listen to, instead of scream-
ing at, reality: opening your mind enough to find what the end of
someone else's sentence sounds like, or listening to a dog until you
discover what is needed instead of imposing yourself in the name of
training.*

— THOMAS DOBUSH, Monk of New Skete
(October 9, 1941–November 7, 1973),
in *Gleanings*, the Journal of New Skete,
Winter 1973

*I love inseeing. Can you imagine with me how glorious it is to insee, for
example, a dog as one passes by. Insee (I don't mean in-spect, which is
only a kind of human gymnastic, by means of which one immediately
comes out again on the other side of the dog, regarding it merely, so to
speak, as a window upon the humanity lying behind it, not that,) —
but to let oneself precisely into the dog's very center, the point from
which it becomes a dog, the place in it where God, as it were, would
have sat down for a moment when the dog was finished, in order to
watch it under the influence of its first embarrassments and inspira-
tions and to know that it was good, that nothing was lacking, that it
could not have been better made . . . Laugh though you may, dear
confidant, if I am to tell you where my all-greatest feeling, my world-
feeling, my earthly bliss was to be found, I must confess to you: it was
to be found time and again, here and there, in such timeless moments
of this divine inseeing.*

— RAINER MARIA RILKE, *New Poems*,
Translated by J. B. Leishman

*But now ask the beasts and let them teach you,
And the birds of the air and let them tell you,
Or speak to the earth and let it inform you,
And let the fish of the sea recount to you.
Which among these does not know that the hand of God has done this,
In whose palm is the life of every living thing,
And the breath of every human being?*

— Job 12:7–10

# Contents

## Sensitivity Exercises

## Starting Off Right

## Standard Obedience Exercises

## Problems

## A Parting Word

# Preface to the Second Edition of *How to Be Your Dog's Best Friend*

Twenty-four years is a long time for any instructional book to go unrevised, but particularly when it applies to a field as dynamic and lively as dog training. Perhaps our biggest reason for delaying such a project until now involved priorities: determining what would be of most help to dog owners and their dogs in the nitty-gritty of their relationships. Experience convinced us that a separate book on puppyhood and a comprehensive set of training videos were of more immediate need, so we applied our energies to those projects.

Still, for a long time now we have wanted to come out with a revised edition of *How to Be Your Dog's Best Friend* that included the most current ideas about training and dog care. Doing so means being faithful to what we have learned through our experience. In the years following the original edition's publication in 1978, New Skete has been privileged to continue its work with dogs and to share in the growing understanding of all things canine. Many of the insights and intuitions that we first articulated have had the chance to age and become more refined. We have also continued to learn much that is new, benefiting from the work of many talented trainers and animal behaviorists, as well as from the generosity and openness of dog owners and friends who have brought their dogs to the monastery. More significant, our own sense of the mystery that a relationship with a dog brings us into has deepened, too. The result for us has been a more mature and comprehensive understanding and love of training and the human-canine relationship.

We hope that this revised edition will inspire and enable you, our readers, to create a more satisfying relationship with your dog, while at the same time discovering the deeper significance and spiritual value of life with your best friend.

# Acknowledgments

All trainers, whether or not they attend schools and clinics themselves, develop a great part of their philosophy and techniques through personal exchanges with others in this field. Many people have helped us through the years, in both our breeding and training programs, and we would like to make special mention of them here.

From the early days of our work with dogs, several veterinarians were in touch with us on a regular basis through our mutual referrals and their active cooperation with us in handling problem behavior, especially Drs. Joel Edwards and David Wolfe, Shaker Veterinary Hospital, Latham, New York; Drs. Eugene and Jean Ceglowski, Rupert Veterinary Clinic, Rupert, Vermont; Dr. George Glanzberg, North Bennington, Vermont; Dr. Robert Sofarelli, Saratoga Springs, New York; and Dr. Charles Kruger, Seattle, Washington.

A great many professional trainers helped us during their visits to New Skete or in many other ways: Joyce and Don Arner of Westmoreland, New York; Jack Godsil of Galesburg, Illinois; Fred Luby of the United States Customs Office in New York City; William Lejewski of the Baltimore, Maryland, Police Department; Sidney Mihls of Englewood Cliffs, New Jersey; and Diane Moorefield of Atlanta, Georgia.

For supplying invaluable source materials dealing with the whole spectrum of canine-human relations, Dr. Benjamin Hart, University of California at Davis; Evelyn Mancuso of Natick, Massachusetts; and Lynn Levo, C.S.J., of the College of St. Rose at Albany, New York, were most generous, as were the following, whose general assistance and encouragement helped make the original book a reality: Elizabeth Ryder and Marian Finke; Helen and Jack Dougherty; Alice Riggs and

Marie Leary; Eva and Ernest Seinfeld; Barbie and Bill Fleischer; Roby and Charles Kaman; Gordon Johnson; Ilse and Tony Govoni; Holly and Paul Carnazza; Roger Donald, Richard McDonough, and Diane Muller of Little, Brown; Jody Milano; the Nuns of New Skete; and our many clients who have entrusted us with the care and training of their dogs.

Since the publication of the original edition of this book, many new friends have been extremely helpful to us in our work. We are especially grateful to Ruth Anderson, Donna Malce, Miriam Barkus, Teresa Van Buren, Dennis McCabe, Cathy Wagoner, and Jeanne Carlson for their friendship and openness in sharing with us insights they have gleaned from working with dogs. Jane Hunter MacMillan has been most generous in supporting our work. Our veterinarian, Tom Wolski, has been a close friend and trusted expert through the years and has had an important role in caring for our dogs. For the new edition of this book, we owe a huge debt of gratitude to John Sann, who came to the rescue with his photographic expertise; to Nicholas Hetko and the Nuns and Companions of New Skete; to our agent, Kate Hartson, whose friendship and professionalism have helped us enormously in our writing endeavors; and to the folks at Little, Brown who have been directly involved with this project, especially Terry Adams, Chika Azuma, and Steve Lamont.

Finally, there is no way that can adequately express our thanks and affection to Helen (Scootie) Sherlock, who throughout these many years of friendship has expended incalculable hours of advice, guidance, and encouragement in every phase of our life and work with dogs here at New Skete.

# An Introduction to Training

# Myths, Mutts, and Monks

It may strike readers as odd to find a book associating monks and dogs. Well, both have been around for a long time. Dogs, we must say, have monks beat by many a century, for according to some legends they even predate humanity.

American Indian myths furnish the most ready examples. For the Kato Indians of California the god Nagaicho, the Great Traveler, took his dog along when he roamed the world creating, sharing his delight in the goodness and variety of his creatures with his little dog. Among the Shawnee of the Algonquin nation who once inhabited the upstate region of New York where our monastery is located, creation was brought about by Kukumthena, the Grandmother, and she, too, is accompanied by a little dog (her grandson tags along as well). Creation in this myth is perpetuated by none other than this mutt, for each day Kukumthena works at weaving a great basket, and when it is completed, the world will end. Fortunately for us, each night the dog unravels her day's work. Those of us who have lost portions of rug, clothing, or furniture to a dog's oral dexterity may never be convinced it could be put to a positive use such as forestalling the end of the world. Still, the myth says a lot about the interrelationship between dogs and humans.

The place of dogs in mythology is by no means limited to North American Indian cultures. It appears to be universal. Greco-Roman literature, for example, features dogs in various roles. Think of Hecate's hounds; the hunting dogs of Diana; and Cerberus, the guardian of Hades. More well known is the tale of Argos, the faithful dog of Odysseus, which is recounted to us by Homer in *The Odyssey*. It is set in the context of Odysseus' return home after a twenty-year absence — ten years fighting at Troy, and the following ten trying to get back to his wife and son. Over the years, everyone comes to believe that Odysseus died in the war, though his wife, Penelope, continues to refuse the amorous advances of various suitors, always believing that she will see her husband again. The irony of the tale is that when Odysseus finally does arrive back home in the guise of a beggar, neither his wife nor his faithful servant recognize him; the only one who does is his old dog, Argos, who has been waiting faithfully for his master to return.

Then there is Asclepius, god of medicine, who as an infant was saved by being suckled by a bitch. As were, of course, Romulus and Remus, founders of the city of Rome (to stretch a point). Egypt's dogs have been depicted prominently in ancient murals, and many dogs have also come to us intact as mummies. Persian mythology features a dog in the account of creation. The Aztec and Mayan civilizations include one as well. Various tribes of Africa, the Maoris of New Zealand, and other Polynesian cultures, along with the venerable Hindu and Buddhist faiths, have all found some key place for a dog in the legends that have been handed down in both oral and literary traditions.

Stories about dogs abound in Zen literature since many Zen monasteries keep dogs, usually outside the gates. The principle koan "Mu" is used to foster enlightenment and involves a paradoxical question about whether a dog has Buddha-nature or not. In another story, a monk is caught in an ironical game of one-upmanship with a dog:

> Once a Zen monk, equipped with his bag for collecting offerings, visited a householder to beg some rice. On the way, the monk was bitten by a dog. The householder asked him this question:
> "When a dragon puts even a piece of cloth over himself, it is said that no evil one will ever dare to attack him. You are wrapped up in a monk's robe, and yet you have been hurt by a dog: why is this so?"
> It is not mentioned what reply was given by the mendicant monk.

And in another, a continuation of the above story, the unpredictable nature of some dogs is equated with reality itself:

As he nurses his wound, the monk goes to his master and is asked still another question.

Master: "All beings are endowed with the Buddha-nature: is this really so?"

Monk: "Yes, it is."

Then pointing to a picture of a dog on the wall, the wise old man asked: "Is this, too, endowed with the Buddha-nature?"

The monk did not know what to say.

Whereupon the answer was given for him. "Look out, the dog bites!"*

We should not shortchange the Judeo-Christian inheritance that many of us share. But in fact the Bible, for reasons we cannot examine here, has only an occasional mention of dogs — for example, "Lazarus' wounds being licked by dogs" or "even the housedogs get the crumbs" in the Gospels. However, the dog reappears in other religious literature, sometimes as a symbol of faithfulness, sometimes as a little detail that lends a warm and human touch to the story of a saint's life. Perhaps the most vivid example of this penetration of folk legend into church tradition is the story of Saint Christopher. Many people will be startled by the way he is pictured in Eastern Christian art. The Menaion, or Book of Calendar Feasts, includes a brief account of each saint's life. We learn from this book that Christopher was a descendant of the Cynocephali, a legendary race of giants with human bodies and canine heads. He is pictured thus in icons. He has the head of a dog but otherwise resembles the conventional image of a martyr, down to the cross in his hand. He was miraculously converted and baptized, and given the name Christopher, which means "Christ-bearer."

Many saints in the Orthodox tradition are called God-bearer or Christ-bearer, a salutary title meaning these saints carry divine qualities within and manifest them in their daily lives. In the West the title was taken literally with regard to Christopher, and the legend subsequently developed in which the man (an unattractive giant still) carried the Christ Child across a flooded stream and was transformed into a handsome brute instead. In Middle Eastern tradition

*D. T. Suzuki, *The Zen Monk's Life* (New York: Olympia Press, 1965), p. 25.

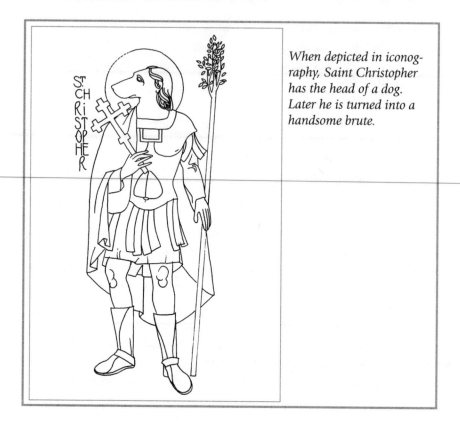

*When depicted in iconography, Saint Christopher has the head of a dog. Later he is turned into a handsome brute.*

he journeyed to Syria to attempt to make an evil pagan king, Dagon by name, see the light. The king was not impressed, even by so formidable a messenger as a dog-faced man. Christopher was imprisoned instead, and in the midst of his martyrdom (he was given the first hot seat on record: Dagon ordered him to be chained to an iron throne and then had a fire built under it — so hot, it is recorded, that both chain and chair melted) he was transformed and received the face of a man.

There is a story, perhaps still told in Romania, where it is thought to have originated, that gives a charming account of how the dog itself was created.* It seems that Saint Peter was taking a stroll in heaven with God when a dog came up. "What's that?" said Saint Peter. God told him it was a dog, adding, "Do you want to know why I made him?" Naturally Peter was interested. "Well, you know how much

*Maria Leach, *God Had a Dog* (New Brunswick, N.J.: Rutgers University Press, 1971).

trouble my brother, the Devil, has caused me . . . how he made me drive Adam and Eve out of Paradise. The poor things nearly starved, so I gave them sheep for meat and warm wool to clothe them. And now that fellow is making a wolf to harass and destroy the sheep! So I have made a dog. He knows how to drive the wolf away. He will guard the flocks. He will guard the possessions of man."

Historically, two groups of monks have been responsible for breeding and training dogs. The canons of Saint Augustine (technically not monks, but members of a religious order) have raised Saint Bernards at their hospice in the Swiss Alps for more than two centuries. The dogs are still bred there, although they no longer perform their well-known rescues of travelers lost in the Pass — airplanes and snowmobiles have limited the need for dogs in that capacity. But occasionally the canons and their dogs still do go out on a search. The famed brandy cask is a myth. It is probably based on the fact that the lost traveler, once found, was usually offered brandy by the Brother who accompanied the search dog. But it was the Brother who carried the brandy, not the dog.

In Tibet quite a different group of monks developed the Lhasa Apso dogs. They raised them in their monasteries and frequently gave them as gifts to nobles. It's interesting to note the disparity in size between these two monastic breeds, as well as that two quite dissimilar groups of monks found working with dogs a fitting monastic occupation. We can attest that raising and training dogs fits into monastic life very well. Dog care takes a lot of labor and affection, and monks usually have both in abundance. On another level, the dog typifies in many ways the mature monk: loyal, steadfast, willing to please, willing to learn.

Monks should not be thought of according to the stereotype that no doubt rests in the back of the minds of many — otherworldly romantics, who with bowed head and folded hands walk in silent procession down medieval cloister walks. Nor does the Friar Tuck image apply, though good nature, healthy appetite, and a bellicose streak will be found to varying degrees in most monks. Actually, the best image to capture what a monk is can be found in the words of the Russian author Dostoyevsky, who remarks in *The Brothers Karamazov* that a true monk is nothing more than what everyone ought to be.

Still, that is certainly debatable: "what everyone ought to be." Obviously, he did not mean that all of us should be celibate. Instead, he was pointing to an attitude of heart that he believed was characteristic of monks. The key to human happiness and fulfillment — for monks and nonmonastics alike — lies in a wholesome spiritual understanding that is supremely rooted in reality. Though monks certainly have no exclusive claim to such an understanding, we do attempt to pursue this in a professional way, passionately searching for the truth of who we are and what life is all about. What we have learned is that for the person who is truly open, the whole of life has the capacity to speak, to become a word leading us to greater wisdom and understanding. We have but to listen. From such a perspective, it is hardly surprising that our dogs have taught us much about ourselves, in many subtle ways showing us how we ought to be, as well as how we ought *not* to be. Because of their association with humans, an association that the stories we mentioned above show to be as old as human consciousness itself, dogs are in a unique position to offer humanity a reflection of itself.

Anyone who knows someone with a pet does not have to search too far to find similarities between the two, in little things, perhaps, in behavior quirks, in outgoing friendliness (or the opposite, suspicious reserve), and even — and often the most amusing — in appearance. Some cartoonists (such as Booth and Price in *The New Yorker*) get a lot of mileage out of the latter. On a deeper level, when we pay close attention, dogs mirror us back to ourselves in unmistakable ways that, if we are open, foster understanding and change. Dogs are guileless and filled with spontaneity: unlike people, they don't deceive. When we take seriously the words they speak to us about ourselves, we stand face-to-face with the truth of the matter. We can easily learn to reflect on these words — they are inscribed on their bodies, in their expressions, in the way they approach and interact with us. There is more raw material for meditation here than in many a spiritual book, which is why we offer our experience with dogs not just for the benefit of your dog but in the hope that you, too, might learn something about yourself through interacting with your dog. A better insight into your dog may suddenly give you a glimpse of your own humanity. Just as important, it often heightens the sense of responsibility we humans have, not just for our fellow creatures but for one another and for all creation.

# 2

# How New Skete
# Went to the Dogs

In Egypt, there is a devoutly religious tribe called the Nuer. The Nuer live near the Nile River and raise cattle for their livelihood. But their cows are more to them than just a source of income. Barns, cow halters, and electric fences are foreign to the Nuer. Instead, they integrate their cows into the total fabric of their daily lives, utilizing them in work, letting them mill around, sleeping near them, and meticulously grooming and bathing them. Each cow has a name and a personal history, known by all the tribe. Daily life is characterized by incessant conversation (or so it seems to an outsider) about the cattle. Each tribesman has plenty of stories to tell about his cows, cows he has owned, or cows he hopes to own. The Nuer are always looking for the "ideal cow." Cows even attend some religious services, and Nuer ritual is full of references to you know what. Nuer religion has been studied extensively and is considered by anthropologists an archetypal primitive religion. The Nuer are, on the whole, physically healthy and psychologically wholesome. They live totally integrated with creatures that are on another level of existence.

Now, what has this to do with the training and breeding programs at New Skete? In some ways our lives at New Skete resemble those of the Nuer, and so we can appreciate many parts of their culture. We, too, consider our animals more than merely a source of money. Each monk is personally responsible for one or two German shepherds, whom he comes to know intimately in the course of his life with them. We structure our monastic life in such a way as to include our dogs on as many different levels as possible, implicitly making room for our dogs not only on a physical level but in our minds as well.

Though this might strike some as peculiar, it is actually entirely in harmony with the ideals of monastic life. Traditionally, monks have had a profound reverence for nature and the animal world because they manifested something essential about the mystery of God. This insight came from working intimately with nature, caring for it and learning its secrets, not just reading about it. Genuine monastic living means living a life without division, looking for God in the soil of each and every moment of daily life, not merely when praying and worshiping. Living in close association with our dogs helps us avoid a temptation that is always present in contemplative life — the temptation to live narcissistically in the dreamy world of ideas. We do not "find" God solely in the interior realm, and when we live our lives as if we did, we fall victim to a dividedness that has profound spiritual consequences. Being responsible for our dogs, living creatures that are needy and vulnerable, helps ground us in reality, forcing us to appreciate the mystery of God in all its length and breadth. There is no conflict here with the ideals of monasticism, only a challenge to live those ideals more fully and integrally. For us, the result has been a mind-expanding experience.

## First Steps

New Skete monastery is in the hills outside Cambridge, New York, near the Vermont border. Early in our community experience, from 1966 to 1969, we had a full-scale farm. At one point or another, goats, chickens, pigs, pheasants, horses, Herefords, Holsteins, and sheep all dotted our landscape. Without realizing it at the time, we were beginning to enter the psychic realm of animals. Our observation of the different farm animals began to educate us in a natural, organic way about animal psychology and behavior. We had a German shepherd dog and had thought about eventually breeding. Meanwhile, the farm animals were an excellent preparation for us. In a sense, training and raising German shepherds is the apex of our long experience with animals. Our farm had to be phased out, since the new property we moved to, high on Two Top Mountain, could not sustain a farm. We then made the decision to enter professional breeding and training.

Brother Thomas Dobush, who died tragically in an automobile accident in 1973, showed a keen interest in breeding and training as early as 1966, when Kyr, our first German shepherd, came with us from our former monastery as we founded New Skete. Kyr was a male,

*Kyr, New Skete's first German shepherd dog, in late 1966.*

a former Seeing Eye student, and a dynamic, intelligent shepherd. After he passed away, the monastery felt so empty without him that we decided to purchase a bitch and plan a litter. From the beginning we studied our breeding and training plans carefully. We acquainted ourselves with any and all information we could find on the subject. We contacted prominent breeders and trainers, asking for advice and counsel. Professionals recognized our sincere interest and desire to learn, and shared their knowledge with us, in time helping us develop a sound breeding program. We owe them a lot, and we shall be forever grateful for their generosity and help.

Our knowledge in dog behavior and training grew naturally out of our experiences with our own dogs. Brother Thomas began training our German shepherds to live in the monastery as a group and maintain quiet and order, important to monastic life. Later our skills appealed to other dog owners, and we began to train other breeds as well. Whenever a new monk entered, he was apprenticed to Brother Thomas and learned training skills. More than merely instructing his apprentices in handling skills and techniques (at which he was an expert), Brother Thomas tried to communicate an intuitive way of dealing with dogs. He emphasized "listening" to the animal and "reading" the dog's reactions. His training and handling skills were thus passed on in an oral tradition that is still alive at New Skete.

*Brother Thomas Dobush (d. 1973) having a ball with Jessie and Bekki, New Skete's first two female German shepherds.*

For more than thirty years we have lived in a communal situation with our German shepherds. The dogs live in a colony of approximately fifteen animals, of different ages and temperaments, and constitute the core of our breeding program. The breeding program itself is modest, with our primary goal being to produce beautiful, healthy dogs of exceptional character and temperament in harmony with the breed standard. We also run a three-to-four-week obedience-training program that specializes in teaching dogs (and by extension, their owners) the standard obedience exercises in a manner that easily integrates into the owners' daily lives. We limit the number of students we take at any one time so that each dog can enjoy personal attention. Though we are familiar with obedience classes, we feel more comfortable with an individual approach, which fits better into our monastic circumstances, in which a quiet, reflective environment must be maintained. This quiet, we believe, helps humans and dogs alike to learn better.

As in the case of neighboring Egyptian tribes wanting to have Nuer-quality cattle and learn Nuer techniques, so too have we had people

*Brother Job Evans
(d. 1994) with Zanta
and one of her pups.
(Photo by Holly
Anderson)*

seeking us out for advice. We feel both honored and humbled by this, and to the extent possible, we always try to be generous and conscientious with our help. We consider each dog we train or breed to be a reflection on our monastery and an indication of what we stand for as trainers and breeders. Additionally, we have tried to make what we have learned available to the broader public. Aside from this book, we have written a second one more specifically devoted to puppyhood, *The Art of Raising a Puppy,*\* and produced a three-part video series titled *Raising Your Dog with the Monks of New Skete.*\*\* We are currently working to expand our website, www.newskete.com, to include resources to help owners with questions relating to their dogs, and we continue to study and keep up with the latest advances in training and dog behavior. In a sense, we are always beginners, and we have found a learner's stance to be beneficial in increasing our knowledge of training and breeding.

---

\*The Monks of New Skete, *The Art of Raising a Puppy* (Boston: Little, Brown, 1991).

\*\*The Monks of New Skete, *Raising Your Dog with the Monks of New Skete* (Atmosphere Entertainment, 1998).

*An aerial view of New Skete monastery. The outdoor pens for dogs are in the upper left, and the puppy kennel building is at the bottom right of the photograph.*

*East side view of New Skete's puppy kennel. These are the secure runs for older pups and the German shepherds being bred. The horizontal purlins are to hold a clear poly cover to protect from winter wind and snow.*

# 3

# What Is a Dog?

Dogs and humans have been together as close companions for the past fourteen thousand years, if not longer. The origin of the domestic dog is still somewhat unclear. We know, for example, that when people and dogs began to live together, the only other animal with comparable dental characteristics was the wolf. The wolf is certainly a proven, but most likely not the only, ancestor. Most authorities believe that the dog is directly descended from the wolf, but others subscribe to a modified theory that teams up the wolf with some other close relative that may have looked more like a dog. The evolution of different breeds is a fascinating study beyond the scope of this book. For those interested in training or just in becoming better friends to their dogs, one fact is important to remember: every dog claims the wolf as an ancestor. Understanding wolf behavior helps you understand your dog.

There is still a great deal of prejudice against the wolf. Today it surfaces when environmentalists and others clash with those who believe that wolves deplete the deer population and attack livestock and even small children. (The fact is, the wolf can aid deer survival by eliminating the weaker members of a herd, and to our knowledge there is no documented case of a wild wolf killing a child.) Since the wolf is a pack animal, it is sociable with its own kind but wary of humans. Many people confuse the hunting habits of the wolf with those of the fox. Though the wolf moves pretty much with the pack, the fox is a solitary hunter. Wolves invariably stay as far away from humans as possible.

Unfortunately, prejudice against the wolf thwarts a possible way of appreciating the dog, since despite their differences, wolf and dog have striking similarities. Both are innately pack-oriented and prefer not to be isolated for long periods of time. Both are hunters who chase down their prey instead of ambushing it like some of their other close relatives. Both are responsive to leadership from an "alpha figure" to whom they look for order and directives. Both use a wide array of body language to communicate within the pack and with outsiders. Some researchers have noted the presence of a kind of altruistic love in wolf packs, the willingness to please another member of the pack without any reward, and the ability to show caring. These last two traits are well known in domestic dogs.

To learn about dogs, learn about wolves. There are a number of exceptional books about *Canis lupus* that can provide you with invaluable background about your dog and his behavior. We present those we have found to be particularly enlightening in the Select Reading List. Reading about wolves in order more fully to understand your dog and his behavior is not going the long way around the mountain. If you reflect on the behavior of wolves, as reported in these books, you will discover an ironic fact: many books on wolves help you to understand and appreciate your dog's behavior better than some of the dog-training manuals currently available. Many of the techniques in this book dovetail with what we know about the dog's close ancestor, the wolf, and, most important, help us avoid unnaturally sentimentalizing our relationship with our dogs.

Today's dogs belong to the family *Canidae,* along with their relatives the wolf, coyote, jackal, and fox. This family of animals is remarkably diverse, but all members are carnivores, all hunt for food (whether alone or in a group), and all are potentially trainable and tend to learn easily. That said, it is important to voice a strong word of caution. With the recent surge in fascination with wolves has come an alarming and potentially disastrous side effect: people rushing out to purchase wolf hybrids as companion animals. It is tragic that at a time when reeducation about wolves has helped overcome age-old prejudices and has allowed us to appreciate them anew, and where controversial and hard-fought reintroduction efforts have restored healthy wolf populations to several select western locations, perhaps the most serious threat to the wild wolf population is the breeding and purchasing of wolf hybrids.

Hybrids will not help perpetuate the species. The unpredictable mix of wolf and dog features often results in a highly dangerous animal (even despite a high level of socialization), and wolf experts are unanimous in discouraging them. People getting a wolf hybrid usually have no idea what they're in for, and for a good reason. For all their similarities to dogs, wolves possess some genetic and behavioral characteristics that in many ways prevent them from adjusting to domestic life. As we have already said, although they are strong, highly intelligent pack animals, wolves have developed an instinct to completely avoid humans, and are naturally adapted for traveling vast distances each day, something that allows them to easily avoid human contact. As such, they pose no danger to humans. When crossed with a dog, however, the wild and domestic ancestries often come into conflict, making for a very dangerous, unstable creature. Having spent fourteen thousand years transforming a wild animal — the wolf — into an animal well suited for human society, it makes little sense to create a highly unpredictable hybrid, particularly when there is so much statistical evidence of their danger. When a biting or mauling incident occurs involving a hybrid, it only adds fuel to the fire against the wolf and makes the good work of wildlife biologists and wolf experts all the more difficult. The best way to care for wolves and their conservation is by not supporting hybrid breeding. Instead, support those groups whose purpose is wolf education, conservation, and the careful reintroduction of the wolf back into the wild, where it belongs.

# 4

# Some Important Terms

## Relationship

Dogs have been bred over the centuries for a wide variety of purposes, but the more grounded these reasons are in a sound, respectful relationship with the human caretaker, the more the unique nature of each dog blossoms. Dogs are essentially social creatures who thrive when directed positively. For this reason our approach emphasizes the social relationship as the single most important factor in your life with your dog.

Developing a healthy relationship with your dog depends foremost on establishing a climate of mutual trust and respect. This begins at adoption and continues throughout your life with your dog. It presumes your willingness to understand a dog as a dog by being humble enough to learn how he is naturally. Crass anthropomorphizing, however well intentioned, makes unfair assumptions about the dog that hinder the relationship and ultimately his full potential.

## The Pack

When we talk about "the pack" in this book, we usually mean the immediate members of the dog's social circle, both human and canine — in short, the dog's owner and those who live with the dog. Sometimes we refer to this group as the "family pack." As previously mentioned, all dogs — from the tiniest Maltese to the Great Dane — have the wolf as an ancestor, and wolves are pack animals. Through domestication, the dog has adapted to life with humans and has adopted us as his new pack. A dog perceives the people he lives with as fellow members of a pack. Once a dog owner understands this, he or she can utilize training methods that intentionally include the dog in the pack while intelligently and humanely lowering the dog in the pecking order.

## The Alpha

Strictly speaking, within every wolf pack there is an alpha pair, the alpha male and the alpha female, who keep order within the male and female branches of the pack, respectively. Though the alpha male is likely to be the single most dominant member of the pack and the one most responsible for its leadership, alpha females have been known to lead packs and they also strongly influence pack activities. The alphas settle disputes between other wolves and may even run interference for younger members of the pack. Depending on the individual pack, the alpha male's role might be that of dictator or guide, or he might adopt either of those roles at different times. All subordinate wolves look to the appropriate alpha wolf for direction. Wolf packs are stable and successful to the extent that the hierarchies are clearly defined.

Domestication of the dog has not nullified this instinct to lead or be led. This becomes a problem whenever an individual dog does not receive proper guidance, through training, and fancies himself to be the leader, or alpha. There should be no question in a dog's mind about who the alpha figure is in his life — you are, whether you are a single owner or a couple. The owner(s) must act as the leader(s) (if two people assume a role of leadership, they should also use the same approach to training). This is not crass domination of a subordinate creature but provides the dog with direction about his proper place and role in the pack.

## Eye Contact

One way the alpha wolf communicates and keeps order within the pack is by making eye contact with the other members. A piercing glance can often stop a fight from developing and settle disagreements. A kindly glance can signify acceptance. We emphasize eye contact in

*Making eye contact and watching the trainer's face are critical to the training process.*

this book because we feel that it is an essential part of the way dog and owner should naturally relate. It can prevent behavioral problems and help stop them if they happen. Gentle looks (not threatening stares) serve to deepen the relationship, communicate acceptance, and establish trust. A hard, penetrating, and sustained look can help stop bad behavior dead in its tracks. It expresses dominance and helps elicit attention and respect. But before the dog can read your eyes, she must look up at you. The techniques in this book encourage the dog to look up, so that eye contact can be made for whatever reason.

## Training

The concept of training as we define it in this book begins when the puppy is born. Training is not simply a set of exercises (heel-sit, sit-stay, down, down-stay, and so on) that a dog must learn when he has reached a certain age. Instead, we approach training holistically as an integrated process that spans the dog's whole lifetime and includes the many different facets of the canine-human relationship. Even though we always insist on the importance of the traditional exercises, we additionally treat many different types of activities as an intrinsic part of training. Training happens on many levels in a dog's life — not just in obedience school. J. Allen Boone puts it well in *Kinship with All Life* when he distinguishes a narrow definition of training from a holistic one:

> If you would understand this secret, you must first understand the distinction between *training* an animal and *educating* one. Trained animals are relatively easy to turn out. All that is required is a book of instructions, a certain amount of bluff and bluster, something to use for threatening and punishing purposes, and of course the animal. Educating an animal, on the other hand, demands keen intelligence, integrity, imagination, and the gentle touch, mentally, vocally, and physically.*

---

*J. Allen Boone, *Kinship with All Life* (New York: Harper and Row, 1954), p. 44.

# 5

# Selecting a Puppy or Older Dog

## What Breed?

If you are considering buying a purebred, you probably already have specific breed preferences. As you get ready to look for a dog, you might review whether the particular breed you like is suitable to your environment and personality. The best way to find out is to talk to someone who has a dog of the breed you are considering. Almost every breed has a national breed club, and these organizations are happy to level with you about the attributes of their breeds. The American Kennel Club can provide you with a listing of national breed organizations.* Some popular books (*Paws to Consider*, by Brian Kilcommons and Sarah Wilson, and *The Right Dog for You*, by Daniel Tortora, are good examples) provide detailed breed descriptions that can help guide you toward a realistic choice for your own unique circumstances.

As for which breed is most trainable, we have our definite opinions, but really no pat answer is possible. In recent years there has been much discussion about the relative intelligence of dogs and inquiries into which breeds rank higher on the scale of canine intelligence.** Unfortunately, it is easy to take the hype surrounding such surveys simplistically, in a way that is misleading and elitist. Although some

*P.O. Box 37902, Raleigh, NC 27627. This information can also be accessed through the AKC website: www.AKC.org.

**Stanley Coren's *The Intelligence of Dogs: A Guide to the Thoughts, Emotions, and Inner Lives of Our Canine Companions* (New York: Bantam Books, 1995) is a particularly balanced and articulate example.

breeds seem to be more trainable than others and thus seem "more intelligent," this perception usually applies to formal obedience training and should not be generalized in an absolute manner. Breeds ranking low in obedience skills, thus earning them the reputation as being "stupid," may simply be more instinctively adapted to other functions and roles. For example, put a beagle in the context of a hunt and it no longer seems to be the stubborn, dim creature that was so difficult in obedience class.

Furthermore, though most experienced trainers will concede that there are definite differences in the trainability and intelligence of various breeds, they also recognize that there are great differences and variations among individual dogs belonging to the same breed. The environment in which a dog is raised — plus the skill of the owner/trainer — often has a profound effect (both positive and negative) on the overall intelligence of an individual dog. Over the years we have worked with many individual examples of breeds that have not ranked high in obedience training yet proved to be eminently trainable because their breeders and owners were extremely sensitive to issues of socialization and good handling. As one might expect, the reverse is also the case. Not every German shepherd is the Einstein that some "shepherd chauvinists" like to think. If truth be told, subtle or not-so-subtle prejudices emerge whenever the intelligence level of different breeds is discussed.

To be useful as a category, canine intelligence needs to be nuanced. As Stanley Coren's work makes clear, the total measure of a dog's intelligence is composed of three different dimensions: *adaptive intelligence*, which refers to learning and problem-solving abilities in which dogs adapt to their environment; *working and obedience intelligence*, the capacity of dogs to respond appropriately to learned commands and to accomplish tasks in the real world; and *instinctive intelligence*, those aspects of a dog's mental makeup that are genetically passed on from generation to generation. Factor into this personality traits such as the desire or willingness to work with the owner, submissiveness, or high levels of dominance, and one realizes that it is difficult, if not impossible, to apply absolutes in determining the intelligence of various breeds. The best we can hope for are general guidelines to help keep our expectations realistic. Any opinions you may hear should be qualified with direct experience with that

*A dramatic example of canine growth. Before you bring your puppy home, make sure you know how big he will get to be.*

breed. Whereas some breeds may have a natural tendency to exhibit certain behavioral traits, the genetic makeup of the individual dog and the specific environment in which he lives will hide or amplify these traits.

## Male or Female?

Just as the potential owner usually has definite breed preferences, he or she often leans toward one sex or the other, sometimes for good reasons, sometimes not. If it is difficult to express any absolute opinions about different breeds and their characteristics, it is harder still to make flat-out statements about male or female characteristics. Within every breed there are males and females who are docile and pliable, and others who are domineering and harder to control. Remember what we said about the alpha male *and* female. In many breeds, there is no appreciable difference between the protection potential of a male and that of a female; in others, the males are protective, but the females tend to wilt in the face of danger. Specialists in different breeds have more detailed information. In general, we counsel novice dog owners to start with a female. Females are usually more resilient, smaller than males, cleaner, and more easily trained at an earlier age. However, if the breeder honestly suggests an individual pup of one sex over the other and seems to have assessed your situation correctly, then it may be a good idea to take the advice. The breeder probably knows more about your chosen breed and this particular litter than you do and is interested in making a good match between you and your puppy. That said, no good breeder would place a puppy simply on the basis of sex. It depends very much on the individual animal.

The prospect of spaying a female should not be a determining factor in which gender you select. For the vast majority of dog owners, neutering and spaying their pets, male and female alike, is highly recommended for both behavioral and medical reasons. Besides preventing unwanted pregnancies at a time when the pet population is skyrocketing to the detriment of our pets, spaying a female prevents the annoyance of regular heat cycles (twice a year for approximately three weeks at a time — plus the undesirable presence of many local male "suitors" at your doorstep), stops pyometra (a uterine infection), and reduces the risk of mammary cancer. Spaying can also have highly desirable behavioral side effects. The fabled weight gain of spayed bitches is due primarily to overeating and lack of exercise (and hence easily preventable), but many dog owners do report that their spayed bitch is mellower and more responsive and retains commands better.

In males, the benefits of neutering are equally compelling. Medically, the neutered dog has no possibility of testicular cancer and a lower incidence rate of prostatic disease and cancerous anal growths. Behaviorally, since many undesirable behaviors in male dogs are sexually based, neutering tends to reduce aggression, roaming, marking objects with urine, and mounting other dogs and human legs, and generally makes the dog calmer and more trainable. Fears (usually on the part of men) that neutering will turn their dogs into wimps or that the dog will be depressed about its inability to have sex are purely anthropomorphic projections. There is nothing romantic involved in canine breeding: it is simply biology, triggered hormonally in the male only when a bitch (if there is one around) comes into estrus, or season. At other times, the male is completely uninterested. Furthermore, neutered males hardly turn into wallflowers. Though generally they tend to be calmer in domestic situations, they are as vigorous and athletic as their intact counterparts and can easily be taught acceptable forms of protectiveness. Throughout the years of our involvement with dogs, we have *never* seen neutering have a negative effect on a dog's behavior. Either the dog's behavior improved or it remained essentially the same. Considering the facts, the only reasonable excuse not to neuter or spay your pet is a serious intention to breed, and thus your criteria for getting a male or female should rest on other factors.

Be realistic. For the vast majority of dog owners it is unwise to choose one sex over the other because of a desire to breed their first dog with a second dog later. Not only may these plans never materialize but most people have little idea of what is involved in breeding and raising puppies. To do it well is demanding and stressful, requiring large amounts of time, serious knowledge of breeds and bloodlines, and possibly high veterinary bills. Don't even think of getting into breeding as a profit-making venture. The money you are likely to spend breeding a litter of pups far exceeds what you'll make from selling the pups. Breeders do it for love of the breed, not money. Well-meaning parents who want to expose their children to the miracle of birth would do far better to rent or buy something like our video series, which includes a dramatic presentation of canine birth and early puppy development. Furthermore, we do not recommend breeding if your dog is not a registered, pedigreed purebred, certified free of genetic disorders such as hip or elbow dysplasia, and of sound temperament. By no means is it a sure bet that whatever puppy you select will turn out to be breedable, for a variety of reasons. Unless you are a knowledgeable, serious breeder, choose your puppy on the basis of his or her merits alone, not as part of a future breeding team.

All dogs bond deeply with their owners, relating in physical, psychological, and emotional ways. The well-trained male displays qualities of dedication and devotion that have inspired awe in many a sensitive human being. Brother Thomas, who began our breeding and training programs at New Skete, once wrote, "We are to listen to a dog until we discover what is needed instead of imposing ourselves in the name of training." Dogs have much to relate to the owner who takes the time to listen.

Nevertheless, people often express a personal preference in selecting male or female, which should be respected. Though there is always the exception to the rule, there are some very prominent characteristics in each sex of the species. As previously mentioned, females are usually smaller and easier to handle, are more affectionate, and train more quickly. On the whole, males tend to be more high-spirited and dominant, more in command of the situation. In spite of their size — from the smallest poodle to the majestic German shepherd — most can accommodate themselves to an apartment or country mansion.

The following is an excerpt from a letter from a doctor and his family who had a New Skete male German shepherd named Azzo. The letter was written after the dog's death, but the essence and spirit of the male of the breed has been captured beautifully.

His devotion was an example, and his restraint. He conversed continuously in silence and displayed a sense of humor and playfulness without which man would be unbearable. His imagination was a joy; his defense of "his" children was so ferocious that he turned robins and paper bags into dragons. Bees were monsters to be attacked, and he consumed more than were good for him. And never in his short life did he have any understanding of his size. His intelligence and selflessness and sensitivity were a challenge. He expected us to be wiser than he was and where reasoning was superior, we could be, but in all else he led us and made us better. His acceptance of our failures and weaknesses made us humble.

## Where to Get a Dog

There are a number of good sources for obtaining a wonderful pet, each with some pros and cons. There are also several possible sources you should avoid.

### Reputable Breeder

An excellent place to get a puppy or older dog is from a reputable breeder. Good breeders have a comprehensive understanding of the breeds they raise and often have committed many years to producing quality representatives in their puppies. They usually have at least one

*A loving mother. Try to see the sire and dam of the litter, if possible.*

of the parents on hand for the client to meet, and tend to be forthright in evaluating whether an individual pup (or for that matter, the breed itself) is suitable, since they have the best interests of the pups in mind. A bad match leaves everyone unhappy. When a good match seems probable, purchasing a pup from a breeder establishes a personal connection between client and breeder that becomes a valuable resource for information and advice in the future. However, not everyone who breeds merits the designation "reputable." It is wise to avoid backyard breeders whose interest in breeding is primarily either financial or sentimental and who have little interest in breed standards. Since you will be paying substantially more for a purebred dog than a randomly bred dog, the breeder should be responsible enough to make a guarantee of temperament and health as well as provide the appropriate health information, a detailed pedigree of at least three generations, and an AKC registration slip. The facilities should be clean, and the breeder should also be affiliated with a local or national club dedicated to understanding and improving the breed. To find a reputable breeder, the best place to start is the AKC website to conduct a Breeder Referral Search. You will be directed to the nearest parent club, which can direct you to reliable breeders in your state or region. Another possibility is to speak with your local veterinarian, who may be able to recommend good breeders whose integrity he or she can attest to.

## Adopting a Dog from the Animal Shelter

An animal shelter is another good source for finding a dog, especially a mixed-breed dog. There are a number of reasons why dogs end up in animal shelters. Some (usually puppies) are there as the result of accidental (and unwanted) breeding; others, because of a family move, divorce, or illness. Still others arrive there because of owner ignorance about the time and commitment involved in caring for a dog. Once the novelty wears off, many owners take the easy way out and simply deposit their dogs in the local animal shelter. Many of these dogs, even if untrained, have the possibility of becoming wonderful pets when given the proper love and training. Shelter dogs are inexpensive and often quite trainable. Owners adopting them provide a humane alternative to the darker fate of euthanasia.

However, there are certain risks a prospective owner takes when adopting a puppy or older dog from a shelter. Owner negligence and incompetence often explain a dog's presence in the shelter, but not always. Sometimes the dogs have manifested problems that, despite the owners' best efforts, were unable to be solved and may require serious time, energy, and expense to rehabilitate. Also, though a young pup ideally should have been with his mother and littermates until seven weeks of age and been properly socialized with human beings, pups may be left at shelters too young (prior to six weeks), well before they've had enough chance to interact with their mother, siblings, and human beings. This can lead to serious behavioral problems later on. Even when he has been socialized well and raised conscientiously, the longer the pup stays in the shelter, the greater the likelihood he may develop fear-avoidance behaviors that are the result of being under-socialized during the fear period (eight to ten weeks of age). Since

there may be no way animal-shelter personnel can adequately socialize the great number of pups they receive, the youngsters may be neglected at a critical time. Thus, if you decide to adopt a puppy from a
shelter, it is usually best to do so when it is between six and eight
weeks of age.

This said, the ASPCA (American Society for the Prevention of Cruelty
to Animals, headquartered in New York City) and independent local
humane societies have made serious efforts to help place dogs responsibly with new owners. Many shelters have introduced behavioral
programs for dogs in the shelter to improve their adoptability: dog
walking, environmental enrichment, and rudimentary training. The
ASPCA's website (www.aspca.org and the related www.petfinder.com)
offers profiles of dogs who are up for adoption throughout the nation.
Just because your local shelter may not have what you are looking for
does not preclude the possibility of finding a suitable pet at another
shelter in your general vicinity.

When you decide to look for a pet from a shelter, the following
information will help you to do so in a more informed way and minimize the chance of taking home a potential problem. First, try to find
out whatever you can about the dog that interests you. Ask shelter personnel about the background of the animal. Sometimes information
is not always available; sometimes it is. Did this particular one live
with a family? Has he been around children, noise, stairs, city life?
Why was he brought to the shelter? When you view the dogs themselves, try to "read" each one individually without staring; glance back
and forth in an unthreatening manner and form a general impres-
sion. Bear in mind that many dogs react aggressively when confronted
with a barrier such as a kennel or cage, so as you pass by, squat down
in front so that you face the cage sideways, not head-on. This posture
is less threatening. Offer the back of your hand on the outside of the
cage for the dog to investigate and smell, and speak to the dog in a
soft, gentle manner. *Never* put your fingers through the bars. Look for
even-tempered approaches: a cheerful greeting — paws on the front
of the kennel with tail wagging, rubbing its body along the wire as if
to solicit petting, or simply a quiet, unthreatened observation of you.
If you are interested in a particular animal, ask to take her for a walk
on leash, in a controlled area. Even then, though, you may not get an
accurate reading of the dog, since she is in a strange environment,

with a strange person. Be patient, and from time to time sit down and let the dog adjust to you. If she is overly aggressive or shy, reconsider taking the dog. Try to remain coldly objective. Don't be swayed by sorrowful eyes, mournful whimpering, or "friskiness" on the part of the animal. If you are considering a puppy, use the guidelines further on in this chapter to help you evaluate temperament; if possible, take along someone knowledgeable about dogs to help you make a selection and provide a second opinion.

One final tip: if you are looking for a dog and find an appealing one who has been spayed or neutered, chances are good that the dog came from a situation in which the owners felt enough responsibility and concern for the animal to have it altered. Above all, don't go to a shelter on a lark. Shelter dogs are every bit as worthy of a serious commitment to love and companionship as purebred dogs. There is never any justification for a cavalier attitude that says, "No problem, he's just a mutt — if things don't work out, I'll simply get rid of him."

## Breed Rescue

Another good source for finding a purebred dog is a breed rescue club. These clubs have come into existence in response to the alarming number of purebred dogs showing up at animal shelters. Generous purebred fanciers who are sincerely committed to their particular breeds open their homes and kennels to individual dogs who, for whatever reason, need a new home. During their stay at the rescue home, the dogs are checked for overall health and evaluated for temperament, basic obedience skills, and house training. Volunteers often work with dogs who need to improve in specific areas to meet the requirements of adoptability. The last thing people working in breed rescue want is an animal to be recycled for adoption again and again.

Checking with a breed rescue club is both a sensible and an exciting choice for someone who is looking for a purebred dog yet may not be able to afford a puppy from a breeder, or prefers a slightly older dog. As is the case whenever adopting a secondhand dog, however, try to find out as much about the dog's background as you can. Pay close attention to the impressions the temporary owner has formed of the dog while in his or her care. Generally the standards of breed rescue clubs on temperament are demanding, particularly regarding aggression, since nothing is gained by placing a poor reflection of the breed into society.

Breed rescue clubs exist for almost every registered breed and can be located by contacting the AKC, or visiting the websites of Canine Connections (www.canine-connections.com) and Pro Dog Networks (www.prodogs.com).

## Where *Not* to Get a Dog*

### *Pet Stores*

If there is one thing professional breeders, trainers, behaviorists, and veterinarians all agree on with respect to dogs, it is this: avoid pet stores when contemplating adopting a puppy. We are not aware of one respected guide to adopting and caring for a puppy written in the past twenty years that recommends pet stores as a place to find a pup, and for very good reasons. Pet stores are profit-making institutions, suitable places for obtaining pet merchandise perhaps, but wholly inappropriate for obtaining a puppy. They acquire the vast majority of their puppies from puppy mills, notorious breeding factories where dogs are bred as livestock, repeatedly and without concern for the health of the dogs, in filthy, inhumane conditions. Socialization is unknown, and vaccinations are rarely given. Puppies are routinely separated from their mothers and littermates prematurely, shipped cross-country to pet stores, often arriving at their destinations weak, dehydrated, and in poor condition. After a superficial sprucing up, stores then often sell them for prices that exceed those of a reputable breeder. With no information on the parents' temperaments or on how the puppy was raised, prospective buyers are making an expensive roll of the dice for the convenience of taking a dog home immediately. Who knows what sickness might be incubating that several weeks later leads to serious veterinary bills, or the behavioral problems that may arise several months down the road because of inadequate socialization? There is, of course, the chance that you may get lucky and get a decent pet, but what is the wisdom in supporting an industry that is overall so abusive to the interests of the dogs?

*See also our discussion of this in *The Art of Raising a Puppy,* pp. 81–83.

Also be aware that pet stores are increasingly recognizing the public-relations disaster stemming from their dealings with puppy mills. Don't be deceived by pet stores that reassure you that their puppies come from "local breeders." Let's get the label correct. These are "backyard breeders" who breed their dogs for profit, with little knowledge of the science of breeding and little real concern for improving the breed (which is obviously why we would not recommend adopting a puppy from such a breeder directly). Even granting their good intentions, we have heard of far too many horror stories to ever recommend such breedings. Furthermore, in our judgment, no ethical, reputable breeder would ever sell his or her puppies to a pet store that allows the pups to be sold to a buyer with nothing other than a valid credit card. Period.

## Choosing a Puppy from a Litter

As breeders, we have had years of experience in placing puppies. We have found that it is not a good idea to allow prospective customers to view a whole litter and pick a puppy on their own. Many times they choose the wrong puppy. All litters have their loners, aggressors, and retreaters. Most people feel that the puppy who immediately breaks out of the litter group and runs up to them, jumping at the fence or barrier, is "the dog for me." We've often heard people say, "I didn't choose him, he chose me — he ran right up to me, and that's how I picked him." The trouble is, often the pup that "runs right up" is the most dominant, and possibly the most prone to behavioral problems in the wrong situation. Clients who are emotional pushovers and are charmed by puppy antics always fall for this approach. Meanwhile, they may ignore other puppies who come up less quickly or who linger for a while. Yet one of them might actually be the right puppy.

When viewing puppies, try to see each individual one alone, in a separate room. This is the only way to get an idea of each personality. Don't try to evaluate a puppy in his litter. It's next to impossible. If you are a novice at this kind of selection process, read up on it before visiting a breeder. The test we use to help us evaluate the personality and temperament of puppies was developed by Jack and Wendy Volhard and is described at length in both our puppy book and video series. *Understanding Your Dog*, by Dr. Michael Fox, and *Behavior*

*Problems in Dogs,* by William E. Campbell, also contain basic puppy-evaluation tests.

Ask the breeder for help in selecting the right puppy. Having observed the pups intimately for close to two months as well as tested them at seven weeks, he or she will have a much more accurate sense of each puppy than you will be able to form in one viewing. For this reason, some breeders insist on making the choice for you. If this is the case, don't be put off by a breeder's thoroughness. He or she might interview you extensively beforehand, to find out what your needs and desires are in a dog. Try to answer these questions honestly and completely. It is an effort for the breeder to do so, and thus it is wise to be a little cautious of breeders who do not. If one does interview you as to your feelings about a dog, it is a sure sign that he or she wants to place the puppies carefully. The breeder will then match a puppy to you and your situation as closely as possible. If you are new at selecting puppies, or if you are in a deadlock position between two puppies you are considering, trust the breeder to make the right choice. The breeder wants to make a good match and will not try to foist a bad or inferior puppy on you. It is in the breeder's best interest to make a good match between clients and puppies. Most breeders have had previous experience in placing puppies, know how to evaluate puppy behavior, and should be willing to share the results of their observations with you.

# 6

# Researching Canine Roots

Newfoundlands invariably like to swim, retrievers like to retrieve, dachshunds like to dig and burrow, and Siberian huskies love to cavort in the snow. Why? That's what they were bred to do. An amazing number of otherwise educated people do not know the original occupation of their breed of dog. In a time when many people are concerned about their "roots," dogs' roots go unexplored. Yet knowing the background of your chosen breed can help you appreciate your dog more fully and even aid in solving behavioral problems.

*Newfoundlands love the water. At a water trial, a Newf tugs a canoe out to a "drowning" victim, at the same time rediscovering her canine roots.*

Dogs are divided into seven basic groups, plus a "miscellaneous" class. There are the sporting breeds, like the spaniels, setters, and retrievers. The toy group includes smaller dogs like Yorkshire terriers, Chihuahuas, or the papillon. Terriers include the Airedale, schnauzer, and fox terriers. The working group, for the most part, comprises larger dogs like the Newfoundland, the rottweiler, and the Doberman pinscher. Hound breeds feature the borzoi, the Afghan, and the droopy-eyed bloodhound. The so-called non-sporting dogs include the much-loved Lhasa Apso, the Dalmatian, and the poodle. The herding group includes the German shepherd dog, the Welsh corgi, and the collie. The miscellaneous class includes breeds not yet fully recognized by the AKC (e.g., Neapolitan mastiff and Beauceron), who are not yet eligible to achieve a championship degree.

These categories do not always help explain the original occupation of the breed. For instance, standard poodles are placed in the non-sporting group, but this doesn't give a hint as to their origins and exceptional retrieving capabilities in the water. It is a very ancient and noble breed, though the public's conception seems to be influenced more by its unusual hairstyle than by its intelligence and high working abilities. It's a similar story with many breeds. Sometimes a considerable amount of research is required to find the raison d'être of a given breed.

The objection that your breed's original occupation no longer exists is no reason to deprive your dog of its genetic heritage. You can provide some kind of modified activity. For instance, Newfoundlands are rarely called upon to rescue drowning victims anymore (except on the coast of France, where they are still so employed), but they might serve as good lifeguards and companions for children on a swimming expedition. A German shepherd living in New Jersey won't have much chance to herd sheep, but it may be of real service as a baby-sitter and protector. The borzoi, a coursing hound, might no longer have the opportunity to chase down wolves, especially in a country where wolves are on the endangered species list, but it might enjoy galloping freely over a wide-open field or golf course. Siberian huskies and other sled dogs were bred selectively to have a strong fore assembly to pull heavily laden sleds. Is it any wonder that many an urban Siberian husky owner has trouble keeping the dog at a strict heel?

It's outside the scope of this book to explore the working history of every breed. But it's a good idea to get a book about your particular breed, and if you have a mixed breed, find out what combination you have — and get two or more books if necessary. Some libraries are stocked with books on the better-known breeds. The more obscure breeds usually have a national club that is willing to send out pamphlets.

Once you find out the background of your dog, spend some time thinking about how you can reach back and enliven his area of interest. Don't be surprised, however, if you get no reaction from your dog. Many Irish setters are no longer interested in pointing, and some cocker spaniels may have absolutely no interest in a woodcock or even in going into the water. In general, breeds that zoom in popularity tend to lose some of their working ability, and their original essence is often diluted as a result of overbreeding and indiscriminate breeding.

On the other hand, searching out your dog's background might give the dog a new lease on life. We once had a Labrador retriever with serious chewing and house-soiling problems. While the dog was with us, we took it into the woods. Immediately, a light went on in the dog's eyes, and when the dog returned from an outing, he was quiet and mellow. Since his master liked to hunt, we suggested having the dog trained in hunting work. Needless to say, the companionship that comes from hunting probably did the dog a world of good, but the chance to express deep-seated instinctual drives might have helped the dog's behavior, too. Chewing and house soiling ceased after the dog became a field dog, and he continued to live in the house.

# 7

# How to Read a Pedigree

Millions of people today are delving into their past, researching their family trees. Nowhere is genealogy currently more popular than in the United States. We are intrigued by seemingly unexplainable talents or tendencies, physical traits, drives, illnesses, or depressions, so we look back to those individuals, living or dead, whom we find melded into the pot of our existence. They provide a strong clue to the mystery of who we are. Though environmental influences play a major role in what we become, they can never increase or decrease what is already within us.

Owning a dog is much like having another member in the family. Here, too, we marvel at all that has gone into the creation of this devoted friend. A mongrel can be a surprise package — a great addition to home life or a mistake that we learn to live with. A purebred often gives us a better idea of what's in store for us. Knowing the history of the breed is an invaluable aid in zeroing in on the individual dog. Knowing the more immediate ancestry is better still.

To most people, pedigrees are as mysterious as the word's etymology. Actually, *pedigree* comes from the French *pied de grue*, which means "crane's foot": the expanding web of ancestral generations resembles the spreading toes of that bird. Just like a family tree, a certified (AKC) or *expanded* (one that includes non-AKC-recognized degrees) dog pedigree provides the ancestral lineage of your puppy. People adopting a pup from a breeder should expect such a pedigree (usually going back to the great-grandparents), and should be wary of breeders who fail to provide one. Be that as it may, the pedigree itself contains a code of registration numbers and earned-degree abbreviations unfamiliar to most owners. Many people who would like to understand their dog's pedigree don't know where to begin breaking this code. A detailed discussion of pedigree analysis and judgments concerning various breedings is beyond the scope of this book; however, it is possible to provide a general description of the information contained in a pedigree as well as questions that it might legitimately raise.

You need several things to make sense of a normal pedigree like the sample expanded pedigree we have provided.* First, you need to have a key to understanding the titles and numbers. Obviously, this would include AKC title abbreviations, but since many working dogs in the United States have been imported or come from German or other foreign ancestry, their titles should be included as well. Prefix titles, those appearing before the dog's name, refer to breed titles and certain specific working titles. For example, the "ch" before the mother's name, Ch. Gondor's Ally CD, indicates that the dam has received her AKC-recognized championship, a certification that her conformation is of high quality vis-à-vis the breed standard. Suffix titles, those appearing after the dog's name, refer to various working titles a dog may have earned: here, the "CD" indicates that she earned her Companion Dog degree (off-leash proficiency in the basic obedience exercises) by competing successfully in several AKC-sponsored matches.

We have provided a list and description of American and German titles and their abbreviations as an appendix. If you are interested in knowing the specifics of what each degree encompasses, we suggest going online to the AKC website for a detailed description of American titles, or the Working Dogs website (www.workingdogs.com) for links to full explanations and rules concerning non-AKC titles, such as Schutzhund degrees. Even though you may not know a specific dog's ancestry firsthand, an expanded pedigree will give you a good idea of the conformation, working ability, and hip status of an individual dog as well as his ancestors.

---

*An expanded pedigree includes all the information provided by an AKC-certified pedigree, plus Special Breed Parent Club degrees and titles as well as foreign degrees. For this reason, it is actually preferable for your breeder to provide you with this instead of an AKC-certified pedigree. The sample here is a pedigree of a fictitious breeding. We are using it solely for the purpose of illustrating and interpreting information contained in a normal pedigree. It does not pretend to showcase real dogs.

Second, it is especially helpful to speak with your breeder or some knowledgeable person involved with the breed firsthand about the pedigree. It makes all the difference in the world when such experts can describe for you dogs they know or have seen, or are well known within the breed, particularly since the dog fancy has grown substantially over the years. There are so many dogs shown and competing these days that it is hard even for the professional to keep track of them, not to mention ordinary dog owners.

As a typical AKC-certified pedigree does not display Special Breed Parent Club degrees and foreign titles, the information it provides the non-fancier may be incomplete. For example, in the sample pedigree "ROM" after the maternal great-grandfather's name indicates Register of Merit, certifying that this stud has produced a significant number of champions. This important information does not appear on the AKC pedigree. Similarly, in the father's name, V Nutz Vom Holtzapple SchH III, V is the German symbol indicating that he was a champion (Victor) in Germany, and "SchH3" shows that he earned his Schutzhund III title, a prestigious and demanding degree demonstrating a dog's mastery in the fields of obedience, protection, and tracking. Schutzhund has three ascending degree levels, I–III; III indicates a very accomplished dog. On an AKC-certified pedigree, that information would not appear and we would have no way of knowing the precise working ability of the dog. This is particularly meaningful in breeds for which many dogs have been imported into this country and are being used as studs, such as the German shepherd dog, rottweiler, Doberman pinscher, and Bouvier des Flandres.

Naturally, the more information a breeder provides on the ancestral background of your pup, the better idea you will have of her potential in whatever areas of training and working you may be interested in. For example, it is helpful to know that the mother also has her CGC (Canine Good Citizen degree), which shows a dog's suitability for being incorporated into the everyday life of the owner. Less formal than the CD degree, its purpose is to demonstrate a dog's mastery of more practical issues of companionship that spring up in everyday life. Seeing this provides an important clue as to how well the pup will acclimate to domestic life. Just because a dog has several working degrees in her ancestral background does not mean that she will fit easily into your life. The activity level of such a dog, ideal for working, may be difficult in a domestic situation. Notice also that in an expanded pedigree the further back in ancestry you go, the more general the information is. This is because of space considerations. In later generations a breeder will indicate information deemed more important.

An AKC-certified pedigree is limited to the following:

- a registration number. Each breed's registration number starts with a different letter. For example, German shepherd registration numbers begin with DL. A date appearing after the registration number (e.g., 10-98) refers to the month and year when the dog's first breeding experience was registered in the *Stud Book Register*. It is not the date of birth. The *Stud Book Register*, published monthly by the AKC, contains the pedigree of a dog that has been used at stud, or a bitch that has whelped a litter for the first time.

- whether the dog or ancestor was a champion (CH before the name indicates recognized conformation excellence within the breed) and/or any other AKC-recognized titles an individual dog may have earned. For example, the mother, Ally, has a CD (Companion Dog) title after her name, indicating that she has mastered the basic obedience exercises off leash by successfully achieving a base score at an AKC obedience trial.

- whether the dog's hips have been certified by the Orthopedic Foundation for Animals (OFA). An OFA number describes the dog's hip status. Dogs with OFA numbers have been certified as being free of hip dysplasia, a major health problem in dogs. For example, OFA32G would mean that the dog was OFA certified at thirty-two months of age with a "good" rating (E=excellent, G=good, F=fair). Certification can be made only for dogs twenty-four months and older. In the sample pedigree, the presence of substantial OFA numbers in this dog's ancestry indicates that the hip background is very good. We should also point out that on the father's side, the German equivalent of a passing hip evaluation is the "a" stamp. Again, the paternal ancestors have strong hip backgrounds.

- whether the dog's elbows have been certified as being free of elbow dysplasia. For example, OFEL28 would mean that the dog's elbows were certified at twenty-eight months of age. German pedigrees do not indicate this.

| PARENTS | GRANDPARENTS |
|---|---|
| **V Nutz vom Holzapple**<br><br>SchH3<br>**DL 641250**<br>10-98 | V Frodo vom Holzapple<br>SchH3, FH, IP3<br>SC 1459281<br>"a" |
| **OFA Hips:** OFA 36G Good<br>**OFA Elbows:** OFEL36<br><br>**DNA Profile:** V62713<br><br>*Call Name: Nutz* | V Helga vom Bilboland<br>SchH3<br>SZ 4056211<br>"a" |
| **CH. Gondor's Ally CD**<br><br>**DL 115120**<br>3-99 | Ch Gondor's Gandalf CDX<br>DL 950126 1-96<br>OFA29G OFEL29 BLK+TN |
| **OFA Hips:** OFA 28G Good<br>**OFA Elbows:** OFEL28<br><br>**DNA Profile:** W 92116 CGC<br><br>*Call Name: Ally* | Rohan's Galadriel CD<br>DL 124121 4-96<br>OFA 26F OFEL26 BLK+TN |

*A sample dog pedigree.*

| GREAT-GRANDPARENTS | GREAT-GREAT-GRANDPARENTS |
|---|---|
| Blitz vom Holtzapple<br>SchH3<br>"a" | Hector vom Holtzapple<br>SchH3 |
| | Heidi vom Claussen<br>SchH1 |
| Magda vom Holtzapple<br>SchH2<br>"a" | Zorro vom Klinegarten<br>SchH3 |
| | Wanda vom Oestricher<br>SchH2 |
| Jesse vom Bilboland<br>SchH3, FH<br>"a" | Meiko von Clusrichter<br>SchH3, FH |
| | Kara vom Bilboland<br>SchH2 |
| Iris vom Klauscastle<br>SchH2<br>"a" | Bernd vom Haus Zero<br>SchH3 |
| | Lessy von Arnerius<br>SchH1 |
| Am & Can Ch Mordor's Raider<br>DL 765120 3-91<br>ROM | GV Ch Mordor's Absolute<br>OFA ROM |
| | Ch Mordor's Winter<br>OFA |
| Ch Gondor's Elfie CD<br>DL 142516<br>OFA hips & elbows | Ch Gondor's Rusty CD<br>OFA |
| | Gondor's Hiya CD<br>OFA |
| Ch Rohan's Aragorn<br>OFA Good H & E | Ch Rohan's Rebel<br>OFA |
| | Arnor's Josie CD<br>OFA |
| Rohan's Thistle<br>OFA Good | Rohan's Onguard<br>OFA |
| | Clearlake's Oksana<br>OFA |

- a dog's color (sometimes). For example, "Blk & Tn" after the OFA number indicates that the dog's color is black and tan.
- a CERF certification, if the dog has one. The Canine Eye Registry Foundation (CERF) works with canine eye specialists who look for evidence of any of a variety of genetic eye diseases. This is important for some breeds such as Labrador retrievers, golden retrievers, and collies that have a particular propensity to such eye problems. The CERF certification means that the dog will be free of specific problems for one year from the date of the examination.
- a DNA profile number, if the dog has one. This is increasingly expected of male stud dogs and guards against fraudulent breeding.

Thus, in reviewing the sample pedigree, there are some conclusions that can be drawn immediately. The pup of this pedigree, a German shepherd from both German and American background, comes from solid working and obedience lines. On his mother's side he has an impressive show background that will affect his general conformation. The mix between German and American ancestry will likely produce a pup who isn't severely angulated, and the fact that he has OFA and "a" stamps extensively on both sides suggests that his hips have a good likelihood of being sound. DNA profiling in the parents attests to the authenticity of the breeding. We can determine that the breeding was an outbreeding, which means that it involved two dogs that were unrelated, as opposed to a linebreeding, which attempts to concentrate the genes of a specific ancestor in order to accentuate positive characteristics of that dog by having that ancestor appear multiple times in the pedigree. It is preferable for linebred ancestors to appear on both the mother's and father's sides of the pedigree. When you see such a linebreeding in a pedigree of a pup you are considering, you would likely wish to question the breeder about the rationale for such a breeding.

Another question provoked by this pedigree would be how adaptable the pup would be to the home environment of the new owner. Although this is an impressive pedigree, it also indicates that this pup has a high potential for working activities. How active does the owner intend to be with this dog? The fact that the mother, Ally, has a CGC is a good sign of the pup's adaptability to domestic life; however, the working potential will need to be exercised since the dog will likely be quite active. If the owner does not meet this need conscientiously, the dog may develop problem behavior related to boredom and poor handling.

Pedigrees point to potential capacities, elements contributing to the mystery of a dog's individuality. Often we find ourselves looking into the eyes of the German shepherds we love and seeing more than the individual dog, looking down the corridors of five, ten generations and more, reaching back to an ancestry of planned genetics that has shaped this animal.

We know, too, that dogs are happiest doing what generations of selection have developed in them. Look for determination and stamina in a bloodhound, a soft mouth on the retriever, speed and vision in the basenji, Afghan, and Irish wolfhound. Give your potential pup an aptitude test, but know that behavior patterns — reaction to external stimuli — are largely inherited. The correct environment, socialization, and TLC are up to you.

# 8

# Where to Find Training

There are a number of methods for training your dog, each requiring a strong commitment on your part to be successful. You can do it yourself with the aid of books and/or instructional videos, you can attend obedience school with your dog, you can work privately at home with a professional trainer, or you can send your dog to an established training school. Though the number of people seeking professional help to train their dogs has risen steadily over the years, by far the most popular method is to do it yourself with or without books and videos. Chances are, if you are reading this book, you might be interested in training your dog this way. This might be the first, or the tenth, book you have read on dog training in your efforts to get help.

Training manuals can be of great value, and it is possible to train your dog alone. For some, the home method is the only possible course because of financial, geographical, or time considerations. But a better method is to combine reading and practice at home with another type of training, either public or private.

## Doing It with Books

If you must train your dog yourself, try this method. Get at least three of the training books suggested in the Select Reading List and one video series, preferably one connected with one of the books. For instance, try reading this book (along with viewing our video series) and then follow up by reading two other books, one by a woman trainer and one by a man. All three books will probably cover the same basic exercises, although fewer training books cover problem behavior, as this one does. Nevertheless, the subtle (or not so subtle!)

nuances in training techniques will broaden your perspective and help you be more flexible toward the specific needs of your dog.

Do not attempt to train your dog with an open book in one hand and a leash in the other. It just doesn't work. Instead, begin by reading one book cover to cover and viewing the companion video. The video is a crucial supplement to the reading, as it lets you see the training exercises fleshed out with a number of different dogs as many times as necessary. Next, read the second book all the way through. Compare the two methods and reflect on them in light of what you know about your own dog. Try to visualize yourself working with your dog so that you become increasingly familiar with the process. If you're in a hurry to begin training as quickly as possible, the third book can be read while you are actually training, but it's best to read all three before you take out the leash and begin a session with your dog.

This is exactly how we began to train at New Skete. Since we had no immediate access to obedience clubs, we took each training book with a grain of salt, reading it critically and applying techniques selectively. Meanwhile, our monastery full of dogs served as an indispensable "lab" for testing and perfecting training techniques.

So if you can't get to a training school or can't leave your dog with a professional to be trained, take heart — you can still train your dog very well. Just be sure to read more than one book, and be aware that the video condenses a month's worth of consistent training into a one- to two-hour-long presentation. Videos are meant to be viewed and reviewed, section by section, as many times as necessary. Don't fall into the trap of thinking that your dog will be trained as quickly as looks to be the case in many a video. Patience, repetition, and realistic expectations are key. Think about the techniques and underlying philosophy in the approaches you use. There are also several excellent periodicals that deal with dog training (see Select Reading List). It might be good to subscribe to one or more — if only during the period of time you are training your dog. Dog training is an evolving field, and no one author has all the answers. Read, reflect, and then begin training.

## Obedience Classes

The obedience school in the park is the second most popular method of training and is a viable option for many dog owners, especially if they take their dogs at a young age. Try to find a trainer who specializes in KPT (kindergarten puppy training) if you have a young dog. Many trainers will ask you to wait to enroll your dog in school until he is six, seven, or eight months old. But this might be too late for some dogs, since behavior patterns are set by then. Don't be put off by the "I'm sorry, your dog is just too young" routine. Inquire specifically about KPT. In the past twenty years KPT has become increasingly recognized as a valuable socializing and preliminary training program for younger puppies, and trainers everywhere acknowledge how such training pays off in the long run. As breeders and trainers, we know that training can begin as early as two or three weeks! There are adult brain waves present in the brain of a three-week-old puppy. KPT trainers are willing to work with your dog the second or third month of his life.

If you have an older dog, you may also wish to enroll in an obedience class. Keep in mind, however, that even though obedience classes remain a convenient and economical way to train dogs, there are a number of qualifying factors to consider. Our experience has been that in the beginning stages of formal obedience training, classes are not as helpful as training your dog yourself or working with a private trainer, since the class setting is inevitably a highly distracting environment. Just imagine fifteen owners, each with a spirited, out-of-control dog at the end of a leash, valiantly trying to pay attention to the trainer's instructions, and you see the difficulty. Ordinarily,

*For the dog who has had basic obedience training, distractions can be of enormous benefit in reinforcing the training.*

the dogs are so focused on one another that it is extremely difficult to recoup their attention without a lot of force and yelling. That can make training unpleasant and unproductive.

Dogs learn best when they can focus their attention completely on the trainer, in a quiet, distraction-free environment. With nothing competing for the dog's attention, it is much easier to bring the dog to a basic understanding of the standard obedience exercises. Then, having reached that point, classes become particularly helpful. For a dog with rudimentary to advanced basic obedience skills, there is real value to working amid a lot of distractions: it can increase the dog's training abilities, socialization, and level of attention. Obedience classes help reinforce the exercises your dog knows (proofing) and prepare her for the many spontaneous distractions of everyday life.

In any event, if you do decide to enroll in a class, make sure you can attend all the sessions. This usually means an eight-to-twelve-week commitment, and it is understood that you will be practicing each day on your own with your dog between classes. Particularly if you are just starting out with your dog, one absence can put your dog behind the others in the class, so approach each class seriously and try to be on time, ready for instruction. You must be attentive and alert. Since the instructor is working with a large number of people, it might not always be possible to repeat information or answer every question. Try to find a small class of five to ten people if you can. Most classes begin with an introductory session without the dogs. This session usually prepares the owners for upcoming classes and includes registration and other paperwork, such as vaccination records. You might be asked to sign a waiver concerning any accident that might involve you and your dog. During the initial session, some trainers may comment philosophically on their ideas of training. Be sure to take notes, and read and reflect on them at home. The initial class is no less important than the sessions with the dogs, and occasionally more so. If a trainer never says anything that hints at an underlying philosophy of training, or gives no indication that his or her program has an ethical base, think twice about that particular instructor.

## Finding a Good Trainer

Finding a good trainer can be a difficult but rewarding search. The first class should give you an idea of what caliber instructor you are dealing with, but by then you might have already plunked down your tuition. It's better to first check up on your area trainers with local training clubs, veterinarians, and friends who have already had their dogs trained. You may also wish to consult Internet resources for nearby trainers who are members of ethically based associations such as the Association of Pet Dog Trainers or the International Association of Canine Professionals or who have been certified by the National Association of Dog Obedience Instructors (NADOI). Next, meet the trainer personally before you go to the first class; however, be sure to make an appointment before you go. You may take your dog, but don't ask for an actual evaluation or training session unless you are willing to pay for individual tutoring. Your initial visit should be short. Introduce yourself and your dog. Explain to the trainer what your needs are and what you hope to accomplish with the training. Try to determine whether the class will meet these expectations; size up the instructor's reaction and level of interest. This initial meeting, however brief, should tell you a lot. Another possibility is to attend a class the instructor is conducting and watch the progress from the sidelines, minus your dog. Observe how the trainer works with both dogs and handlers. Does he or she encourage the handlers? How does he or she interact with the dogs? Does he or she appear unduly harsh and punitive? Does he or she stop to answer questions? Are his or her instructions loud, clear, easily understood? Do the dogs and handlers look bored? Does he or she talk too fast?

Beware of instructors who trumpet the number of dogs they have trained, who name-drop clients who are famous stars, or who try to handle large classes of more than twenty dogs alone. A little humility is important to look for in a trainer. A good instructor will have at least one apprentice or assistant in a large class. If an instructor seems too physical in managing dogs, ask about it. Training methods do occasionally include physicality and force, but excesses should never be tolerated. Don't overreact if the instructor applies physical discipline in cases of aggression or extreme disobedience, especially if a handler cannot or will not control a dog. The instructor is responsible for the safety of the handlers and dogs in the class. If you do not understand something the instructor does, ask about it.

There are some instances when you should simply quit class and walk out. If an instructor "hangs" a dog (except as an absolutely last resort to prevent an attacking dog from seriously biting the instructor), swirls a dog around on the end of a leash, kicks a dog (except to stop a real dogfight), insults a handler, consistently refuses to answer questions, or derides the dogs, quit. But don't jump to conclusions, and ask for clarification before you take action.

## Personal Attention

Don't expect tons of personal attention in dog obedience school. If you are lucky enough to find an instructor who insists on small classes, you may get a lot of personal help. But in most large classes with a set time limit, the instructor simply can't stop to take five or ten minutes with each person. You and your dog are a team that is working in a class, but to reap a real reward, you have to plan on consistent daily work on your own at home. Our clients report that lack of individual attention is the biggest single drawback of park-type obedience courses. Clients needing or desiring much counseling should not expect to get it in a large class. Some instructors, while quite skilled in teaching basic exercises, have little or no experience in diagnosing more complicated canine behavior and can sometimes hand out bad remedies.

## Flunking Out of Obedience School

Don't be discouraged if you "flunk out." You probably won't, but if you do, take it in stride. It certainly does not mean that you or your dog is untrainable. Like many children, some dogs simply cannot take the structured school approach, or they may need individual training. Go to a trainer who will work with your dog alone. Or begin again by seriously studying training manuals, teaching your dog yourself. Ten percent of the dogs we see at New Skete are obedience-class dropouts. Almost all respond to a more individual, concentrated approach.

## Private Training

Private training in which the trainer comes to your house is a third method of training. Although more expensive than a local obedience class, the personalized attention you and your dog receive are often well worth it. In general, it is easier to tailor private training to an owner's specific needs and problems, and at a time that is convenient for you.

With a qualified private trainer you have the advantage of his or her undivided attention during the session, and the training can proceed more naturally at your own pace. Not only are you able to observe firsthand his or her techniques, but, more important, the skilled private trainer is continually helping you learn how to train your dog. Remember, that's the real point. You need to become comfortable working with your dog in real life, and this is more easily achieved under the tutelage of a professional who can steadily help you and your dog overcome the challenges of your specific living environment. This is particularly so in regard to setting up and dealing with distracting situations that have been difficult for you to deal with and that you need to be able to handle in everyday life.

Follow the guidelines for finding a trainer mentioned above and expect a qualified private trainer to be able to provide you with credible professional references from veterinarians and humane societies, as well as references from satisfied clients. Ask him or her if it is possible to observe a session, and assess how he or she communicates not only with the dog but with the client as well.

## Individual Counseling and Training

Resident training is another method of training that has become more and more popular over the past twenty years. Leaving your dog with a trainer can often be combined with a long vacation or other absence when you would have to board your dog anyway. Make sure you are clear about the trainer's facilities. Although many kennels are not able to provide you with a tour of the inside kennel space because of insurance and health considerations (e.g., viruses can be inadvertently imported by customers), they can give you a detailed description of the kennel facility, the size of the individual dog pens, and a daily schedule for care of your dog and should be willing to show you the kennel from the outside. Do the outside kennels have adequate shade? Ask to see the trainer work with an experienced dog, and inquire specifically about what your dog will learn.

Many owners wonder about the transfer-of-training problem. "If I leave my dog with another trainer, will he mind me when he gets home?" is a legitimate question. The answer is usually yes — if you follow up faithfully on what the trainer does with your dog. Most trainers give you literature to read while your dog is away. It pays to read it.

When you come for your dog, the trainer should provide a comprehensive report and explanation of the training, including practical demonstration for you in a situation in which your dog does not know you are there. This can be helpful in two ways. First, it brings to life the literature you have read and lets you observe someone else working with your dog, providing an invaluable mirror. If you are so inclined, ask the trainer whether you may videotape the demonstration for your use as a future reference. Second, since the dog will be anxious to be reunited with you, such a demonstration allows you to observe your dog in a normal session, without the complicating stress of the dog having to perform on the heels of a happy reunion. Once the dog settles down after the reunion, you should be able to practice for a time under the watchful guidance of the trainer. Most trainers offer to have you back for refresher sessions later, if you need them.

Many owners find this method of training a relief and a pleasure. Owners who do not have the time or skill to train through books or in a class may find this method a good alternative. Owners who have flunked out of obedience school, or cannot find an acceptable group instructor, might opt for individual training. Physically handicapped or elderly persons often appreciate individual training, which spares them the bulk of the initial highly physical training. Finally, problem dogs are sometimes more responsive to training away from their environment. But remember: no matter how well your dog is trained in such a program, you still have to practice regularly with your dog for several weeks following his return to ensure a smooth transfer of the obedience. The difference is that now it will be so much easier for you to provide that continuity.

All in all, individual training is a good alternative to class obedience training, and many trainers prefer it to class work. It is usually more expensive, because of labor and board costs, but many owners find it worthwhile.

## Dog-Owner Counseling

Deep-seated canine behavioral problems, however, cannot be solved simply by attending obedience classes. Though your dog might become expert with the heel, sit, stay, and lie down commands, the living-room rug might still get chewed, the backyard excavated, or the neighbor's chickens chased and killed. Especially in the case of aggressive behavior, try to get individual attention and dog-owner counseling.

If a good obedience-class instructor is hard to find, a good dog-owner counselor is even more elusive. Many owners turn to their local veterinarians. This route might be helpful if the veterinarian is skilled in canine behavior and has the time to talk. But many simply cannot take the time to diagnose intricate behavioral problems. Some vets keep on file the names of dog trainers who specialize in dog-owner counseling. You may have to call several area vets before you find one who knows where to find such a counselor. At New Skete we are experienced in this kind of training, along with other types of instruction. Dog-owner counseling is a growing, evolving field. It takes time and patience. Not every trainer is interested in it or capable of it. Fortunately, the number of people who are training themselves in dog-owner counseling is growing. Many veterinarians are becoming interested in problem behavior. Ten years ago it was nearly impossible to find a professional trainer or veterinarian who would sit down with you and discuss, in detail, why your dog bites, chews, digs, whines, kills other animals, house-soils, or chases cars. Advice on such complicated matters was obtained on the way out the door of the veterinarian's office, or over the coffee table from friends. Ten years from now, individual dog-owner counseling might be the rule, not the exception.

Bear in mind, too, that trainers are not oracles or gods. They come up against problems that challenge and baffle them. They meet canines they can't understand. Hopefully, they have someone they themselves can turn to for help, as J. Allen Boone did in *Kinship with All Life*. He went to visit Mojave Dan, a wise old desert hermit who lived with a colony of dogs and burros. He asked the hermit to help him understand his dog and get at the truth of the animal. The sage thought for a while and then answered, "There's facts about dogs, and there's opinions about them. The dogs have the facts, and the humans have the opinions. If you want facts about a dog, always get them straight from the dog. If you want opinions, get them from humans."

# 9

# The Concept of
# Praise in Dog Training

We asked a number of dog owners what they did to praise their dogs. Here are some of their replies:

"Treats. He loves them. Then I pet him all over the head and shoulders."

"I give my dog a good rubdown. She rolls over on the floor, and we have a great time."

"You can pet a dog all you want, but nothing matches a good bone."

"I talk 'baby talk' to my dog, then I pat his head while he's sitting."

"My dog nudges me all the time for praise, so I wind up with a hand stroking him ninety-nine percent of the time. We look like a couple going steady."

"I never give any praise. Duke comes over and gets it! I don't know how many times I've had my cup of coffee spilled in my lap in the morning when he nudges my elbow, or how many times I've been unable to read the evening paper because he's bothering me for attention."

"Is praise really necessary? I mean, a good meal, a warm place to sleep, isn't that all they need?"

"She's only three months old. I know she needs a lot of encouragement and praise, but if I touch her, she breaks down and wets all over the place."

"The children play with Yalk all day, but he wants another kind of attention, which he gets from me. I don't know how to explain it, but we talk — and it's different from the kids. He'll be outside playing all day, but after dinner, he'll come in very quietly and we have a talk."

Several themes run through these replies: techniques of physical and verbal praise, limiting praise to food treats, dogs who demand constant praise, owners who do not realize the value of praise, and dogs who are easy or difficult to praise because of behavioral or genetic weaknesses. Surprisingly, we find that people often misunderstand what it means to praise their dogs: either they look at it merely as a useful tool or they feel uptight with the whole idea. What is the proper place of praise?

Let's begin by saying that praise is absolutely necessary. It is the cornerstone of any successful dog-owner relationship. It is not a frill, an attached reward for good behavior: unfortunately, this is the most frequent use of praise. This is part of the misunderstanding of praise — people use it as a bribe to extract good behavior from the dog. Dogs see through such calculated insincerity and grow deaf to it, leaving the owner to search for the next technique that will solve this or that behavioral problem. But in a healthy dog-owner relationship, praise is an entirely natural reaction, an instinctive attitude toward the dog that is characteristic of the responsible owner. It is a way of relating that the dog picks up and responds to wholeheartedly. The most common mistake is to consider praise as simply a reward. Rewards do have a place in dog training, but they are not the essence of praise.

## Physical and Verbal Praise

The concept of praise is twofold. Praising a dog is a physical and verbal involvement with the animal that is influenced by the specific personality of each dog. It is a delicate matter to combine the two in the right proportion. Each dog needs and desires a different type of praise for different actions. Most owners understand that physical praise means petting their dogs, but only a few extend any kind of physical contact beyond the head and shoulder regions (see chapter 24 on massage). Others pound on their dogs, and some pet a dog the same way they stroke a cat. Dogs generally like body contact but do not appreciate slapping, heavy pounding, or pulling.

Physical praise needs to be meted out according to the situation and according to the dog. We once had a client who was training her lively Labrador retriever with us. Each time the dog sank into the automatic sit, which is a normal part of the heeling process, the client would explode into lavish praise. The dog would then break the sit, dance around, jump up, and generally go berserk. "He didn't win an Olympic medal," we remarked. "Why not tone down the level of praise so that your dog can handle it emotionally?" A simple adjustment, but the effect was dramatic.

On the other hand, we recall another client, a somewhat quiet and withdrawn gentleman who was working with his aloof shar-pei. Though the dog would come when called, she did so with such lackluster enthusiasm that the man felt embarrassed. We advised him to be more verbally animated with his dog and to vigorously stroke the dog's neck and side. "But that's just not me," he protested. We made little progress until his daughter happened to come by with his grandchild. Immediately his eyes lit up and he greeted her enthusiastically, extending his arms and speaking to her in a playful manner. The man seemed astonished when we pointed this out, and suddenly he was able to make the connection. In time, his dog started coming out of herself, displaying more animation and affection and less independence, despite the fact that she was a shar-pei.

Try to match physical praise to the situation and to the individual dog. Avoid overloading active, lively dogs with too much physical praise when doing so is going to be counterproductive. At the same time let it become the natural response when a dog needs to be brought out of herself. One member of our community described how he matches praise to one situation: "Getting up in the morning, my dog stretches, and comes over to the bed just about the time I put my feet on the floor. I vigorously fluff up her mane and ears, ending with stroking and patting her about her muzzle."

Dogs need verbal praise as much as (if not more than) physical praise. It's difficult to understand what intense verbal praise can mean to a dog until you see it in action. In our early days of dog study, we read about "animation" and the importance of varying voice tonality in relating with canines. But no book conveyed what we experienced when we heard Helen (Scootie) Sherlock of Caralon Kennels relating with her dogs. She usually has a large number of German shepherd dogs surrounding her, all seeking attention. Scootie is extremely verbal to begin with, and she has a special lingo for her dogs.

In the midst of morning chores, Scootie cannot possibly stop to relate physically with each dog. But as she weaves about, feeding, watering, cleaning up the kennel, she manages to include each dog. Each one feels personally noticed. They immediately become animated and focus their attention on their owner. Since this is their first reaction of the day, they start off with an accepting, willing-to-please attitude toward Scootie that makes the rest of the day easier for everyone.

Notice her technique: she says a few introductory words, then inserts the dog's name in a high-pitched cadence. If the dog doesn't realize in advance that he is being addressed, the inclusion of his name makes it a certainty. It's often difficult to get your dog's attention if you always use the call name first. Try a few happy introductory words.

Scootie uses common slang and CB-radio lingo because she is comfortable with it and because it delivers strong staccato sounds with clear tonal contrasts in short syllables. We've found that many people are verbally inhibited and find it difficult to loosen up and talk to a dog. They find it uncouth, babyish, or demeaning. We've seen some starchy types muffle their embarrassed smiles when they hear us break into a song to one of our dogs, often a personalized jingle about the dog's good or bad qualities. Our Sister Anne has a saying that comes out of our experience with our dogs, "every dog has a song," and repeatedly we have seen how such jingles enhance the relationship. The dogs "read" the expression and tone contained in the melody and respond joyfully. This is not unique to us. All great trainers animate their dogs by talking in happy, peppy tones, employing key affectionate phrases, and using the dog's name frequently. Trainers from as varied training backgrounds and philosophies as Diane Bauman, David Dikeman, John Ross, Jack and Wendy Volhard, Ian Dunbar, Brian Kilcommons, Donna Malce, and Sheila Booth (just to scratch the surface) make verbal praise an organic part of their training methods and their daily life with their dogs. Whether they use food, clickers, toys, petting, or even nothing else in conjunction with the training process, good trainers manifest warm, sincere praise as a natural part of their interaction with their dogs.

Another tip: your verbal praise need not be limited to "good dog!" Vary your praise. By avoiding a predictable and boring expression, you can make up a whole list of praise words and phrases that communicate enthusiasm and get your dog's eyes focused on you and his tail moving vigorously. "Yes!" said with a high-pitched intensity is particularly effective as praise when your dog gets something right. "That's my girl [guy]," "super," "better," and "outtasight!" can also be used in an entirely natural way to add attention and focus. Depending on the intensity of your dog's response, you may have to fine-tune the pitch of your voice; practice and find out what elicits the most favorable response.

## To Treat or Not to Treat

Confusion over whether praise is just a reward for good behavior or an entire attitude toward the dog results in the *substitution* of food treats for physical and verbal praise. Food treats are an extremely effective motivator to help some dogs learn. They also help maintain attention and reinforce a correct response. However, they are not meant to replace sincere verbal and physical praise, which should characterize the whole of the relationship with a dog. These should always be your primary means of expressing affection. Good trainers who use treats in their programs equally advocate verbal and physical praise as vital elements in training, so it is never a question of either/or.

Used in a system of conditioning during formal obedience training, treats can be powerful positive reinforcers that enhance the speed with which the dog learns an exercise as well as his willingness to perform it. However, once the skill is learned, you should gradually wean the dog from the treat, giving it more and more intermittently. This maintains the dog's level of interest (since she's hoping that a treat will come), until finally you are reinforcing with warm praise alone, and saves you from perpetually having to carry around treats and feed aprons. Using treats is outlawed in the obedience ring; anyway, it makes sense to train your dog to obey willingly for the sake of the relationship.

Apart from formal training, there is nothing wrong with using a periodic treat on an informal basis (for example, a biscuit at night). Here at New Skete a monk will often give his dog one of our dog biscuits before retiring, but only after the dog complies to a sit or some other string of commands. Dogs that beg, jump on their handlers, steal, get up on counters, or in any way display bad behavior in connection with food should be put on a program of regular obedience training (with emphasis on the down-stay), a strict schedule that limits their freedom (e.g., alternating between periods with the owner, in the crate, in an outdoor pen, and back with the owner again), and set-up situations that allow the dog to "win" by earning warm praise and perhaps a treat. Neurotic dogs play games with their owners by nudging them for treats, getting the treat box and dropping it in the owner's lap, and refusing to obey unless treated. Don't cave in to "treat games"!

One of our cardinal rules in handling behavioral transgressors is to limit the dog's freedom by adding positive structure to her life to help place the dog in proper relationship with the owner. This approach has to be grounded in regular obedience training. If treats are used, they should be used judiciously, only to motivate the dog toward appropriate behavior, and always accompanied by warm praise. *Never* use treats to reward behavior you have not specifically asked for. *Never* award treats from a false sentimentality that feels sorry for the dog or thinks he is being cute. That is harmful to the dog and is simply pandering to your own need to feel good about yourself. Though we are not against them, treats should always reinforce the dominant leadership of the human in the dog-human relationship and should always be accompanied by praise. The majority of serious problem-dog situations involve an unhealthy and unbalanced relationship between owner and dog, so if the dog understands the basic exercises and is still taking advantage of the owner, we suggest suspending treats until the dog's behavior is acceptable and the relationship more clearly balanced.

Another suggestion: instead of commercial dog treats, try making your own. A relatively inexpensive and effective motivator is a hot dog. We slice it into very small portions and then use them as needed in a session. Most dogs also like liver sausage (rolled into small tidbits), chicken, ham pieces, and sometimes cheese kept in a training apron. For a treat outside of training sessions on a hot day, try an ice cube. Some gourmet types will simply let it melt on the floor, but many dogs love the crunchiness and coldness of an ice cube. It helps prevent dehydration. It's a treat that's considerably less expensive than commercial products, and your supply is usually unlimited.

## Antipraise Owners

Some owners do not realize the value of praise or may even have a deep-seated prejudice against it. Occasionally we have a puritanical client who declares categorically, "Dogs do not need praise. It spoils them and makes them take advantage of their masters."

One man who felt this way wanted to know if we could train his cocker spaniel to the down command in response to a cough. The gentleman explained that he wanted complete control over his dog and didn't want to have to bother giving the cocker a command, but he thought an "ahem-type" cough should do the trick. He then demonstrated by clearing his throat suggestively. He never praised the dog verbally or physically. "My family had plenty of dogs," he explained, "and none of them needed to be hugged every two seconds." The children in the family sat rigid throughout the interview, contributing little. The wife contradicted the husband at one point and he shot a silencing glance at her, clearing his throat in the same suggestive manner! We explained that dogs need vocal commands and hand signals in order to understand clearly what is asked of them. The cough idea was not possible, especially in a dog-owner relationship that was already faulty and plagued with chewing and house-soiling problems. Luckily for dogs, it's rare to find this type of autocratic owner.

## Behavioral and Genetic Difficulties

Some dogs have behavioral or genetic troubles that make praise difficult. We mentioned earlier the dog that plays the "treat game," cajoling the owner into dishing out tidbits. Some leader-type dogs demand affection and praise constantly, nudging their owners, jumping, yodeling, and making life difficult until they get it. This type of dog always seems to want to be the center of attention. The minute the spotlight shifts elsewhere, as often happens when company arrives or if the owner is on the phone, this dog begins the "attention game." Visitors are also nudged and pummeled, sometimes in the genital areas, until they give in and pet the dog. The dog refuses to lie down and, if isolated, he does damage. Often the owner has emotionally overloaded the dog by going along with the attention game. The solution should include a consistent program of basic obedience, at least to the come, sit, and stay level, and reordering the relationship so that praise is given only in response to an obeyed command like sit, come, or stay.

Genetic faults often complicate praise giving. Submissive urination in puppies and occasionally in older dogs often happens in response to physical praise. Ignore this type of wetting, and try to shift to lighter verbal praise rather than physical praise until the dog develops more bladder control. Do not discipline submissive urination. It is not the same as house soiling.

Other dogs come from bloodlines that are so hyperactive that praise elicits excessive shaking and nervousness. Again, try to develop quiet and peaceful verbal praise with this type of dog. Such dogs are not hard of hearing and can respond well to praise given almost at a whisper. Some dogs practically have a nervous breakdown when physically praised. Warm, properly inflected verbal praise usually does not elicit the same response. Physical praise for nervous dogs should be given only when the dog has responded to the sit command and is anchored. Jumping up can then be controlled.

Praise, then, is more than treats, more than an occasional physical pat, and more than a reward for good actions. Praise is an attitude, a stance. Dogs who live in an atmosphere of praise come to love the human voice. They are more trusting and accepting. They are approachable by strangers but not demanding. Dogs confident of praise from their owners do not live on the edge of an emotional abyss, always seeking out attention and sulking when they do not get it. If praise is part of your attitude toward your dog, you have a rich and exciting relationship ahead.

# 10

# Discipline: The Taboo Topic

Some dog-training books never mention discipline, whether in the context of training or for outright bad behavior. Yet questions about it are frequent in our consultations. Owners run the gamut of emotions and responses in their attitudes toward correcting their dogs. Here are some sample quotes we've collected over the years when we asked the questions "How do you correct your dog?" and "Do you discipline your dog for bad behavior, and if so, how?"

"I hit him on the rump with a rolled-up newspaper. Sometimes I have to chase him. He knows when he's done wrong."

"I yell and say, 'No, no,' and she slinks and hangs her head. But then she messes in the same place the very next day."

"When I disciplined the dog, the children would scream and cry, so I gave up. I didn't want the kids to think I was hurting the dog."

"I tell her to stop it and jerk the leash, but it doesn't seem to make a dent."

"Even if I raise my hand to smack Queenie, she bares her teeth. It's like living with Hitler."

"I never hit my children, and I never hit my dog."

"One dog we had — we beat on him pretty hard. We broke his spirit, and he took off on us. I don't want that to happen again."

"We discipline by hitting the dog over the head with a stick. It works."

"From puppyhood on, I punished my dog by smacking her rear end. Then I went to obedience school when she was about eight months old. The instructor said to give the dog a light tap on the rear end to get her to sit. This didn't work with my dog. She would urinate when I touched her rear end. She thought she was being punished. I had to find another method of teaching the sit, and I fell behind in the class."

"I simply don't believe in discipline, physical or otherwise. Yes, I am my dog's maid. I don't like the arrangement, but I've seen other dogs who were hit, and they always look sad."

"My dog cheats, steals, craps in the house, and has bitten three people. You tell me to discipline him. Okay, where do I start?"

"If I even look at Buffy cross-eyed, that's enough discipline for her."

"I lost control one day and smacked Butch under the chin for stealing a rib roast. He hasn't stolen anything for three months. I think I got through to him."

"On the rear end, with a hair brush."

"A good kick usually does the trick."

"For housebreaking, I rub his nose in it, and for chewing, I cram it down his throat."

"There's a great deal of inconsistency in my family. Some are pacifists who would sooner die than hit the dog. Others are bullies who would torture the thing if they got the chance."

"A good night outside usually shapes them up. Out into the cold!"

"Let me warn you: if a trainer ever hit my dog, I'd kill him."

"I say, 'No,' King looks away, and then I end it. I can't do anything else once I look at his face."

"I want to understand when and how to discipline my dog, but the training books talk about everything but that, and I feel I might do something wrong. What exactly do you do?"

"For the life of me, I don't know why Prince won't stop chewing. I beat him every night!"

Perhaps one reason dog discipline is shrouded in mystery is that most owners are simply afraid of discussing the subject. Many find the idea of disciplining their dogs threatening and one that too easily invites feelings of guilt. Since dogs can't talk back to us, it is easy for well-meaning owners to project all sorts of negative, human feelings onto their dogs when occasions of discipline arise. Add to this the periodic flurry of horror stories about dog-beaters, irresponsible trainers, and gross incidents of generally inhumane treatment of dogs, and it is little wonder that simple and effective discipline for disobedience and bad behavior then gets confused with cruelty.

And it doesn't stop there. Further complicating the issue is the division of opinion within the training profession itself. Among dog trainers and animal behaviorists today, perhaps there is no more contentious issue than the place of discipline (and by extension, punishment) in the relationship with a dog. Opinions about its appropriateness run the gamut from harsh, physical corrections, to electronic corrections, to physical corrections of a lesser nature (simple leash corrections or verbal reprimands), to no corrections whatsoever. Owners are understandably perplexed about what exactly to do when a dog misbehaves. Who's right, anyway? Which approach is effective and truly compassionate; which has the dog's best interest at heart?

One current line of thinking suggests never using corrections at all. For example, an increasing number of training methods market themselves on the premise that the only humane way to train a dog is with purely positive reinforcement, that is, rewarding desired behavior with food and praise while ignoring bad behavior. This perspective deems any sort of discipline or correction to be abusive. Yet others wonder how realistic and humane such an approach really is for the majority of dogs.

Though it is understandable that most owners would prefer to use no force whatsoever if they could change an undesirable behavior in their dogs, it is fair to ask whether such an approach is reasonably possible and in harmony with the natural dynamics of pack existence.

In a wolf pack, discipline (a penetrating look, a growl, pinning down, or a nip) is given regularly by the alpha, and pack members learn limits and expectations very quickly. The same is true in a litter of puppies: the mother uses just enough discipline to get her point across. She has no scruples about slamming an offending pup to the ground, and (wonder of wonders) the pup learns quickly. We do our dogs a disservice if we do not include as part of our relationships with them equivalent limits and appropriate social training. The question is, how do we do so in a skillful and precise way, using a consistent and fair approach to discipline? Should owners feel guilty for using appropriate discipline when their dogs misbehave? Given the nature of dogs, discipline seems to be as much a part of a relationship as companionship, play, work, and affection.

Though training and prevention are essential tools to raising a dog properly and are always preferable to remedial action, one cannot possibly foresee all the situations that arise with a dog. Your dog's respect for you and his acceptance of your role as leader will shepherd you through many a potential problem if he understands that not paying attention to you in a given situation will have serious, unpleasant consequences. A record of fair discipline makes your role as alpha believable. Let's not mince words: through the years we have seen many, many

*The mother teaches pups how to play safely and disciplines them when they cross the bounds of respect.*

good dogs become problem dogs simply because owners were unwilling to teach and enforce limits. Part of this means disciplining dogs when they behave inappropriately. If one timely, effective correction is able to teach your young pooch never to growl at a stranger he is being introduced to (and we have seen this happen often), we feel no need to apologize for exercising such authority. It may ultimately save a dog's life.

## A Sane Approach

First, let's be clear about exactly what we mean by discipline. People are often uncomfortable with the word *discipline* because they immediately think exclusively of punishment, which is never emotionally neutral. But discipline has a far broader scope of meanings. The word has a definite positive connotation as well, especially when we look at its etymology. *Discipline* is related to the word *disciple*, "one who follows," which in turn comes from the Latin word *discere*, which means "to learn." This implies that good discipline flows from good teaching and good leadership; responsible owners, like good teachers and leaders, tailor their discipline to the needs of their own dogs.

Good discipline takes place on a number of different levels. In a program of regular obedience training, discipline occurs naturally as you begin to refine, repeat, and reinforce the basic exercises already learned in a general way. In such a context, "serious punishment" is rare. Though any "aversive" (no matter how slight) can be and is technically described as "punishment" in behavioral science, the reality is that such so-called punishment also covers a broad spectrum. There is a vast difference between a clipped "nah" and a sharp cuff under the chin. Strictly speaking, both are positive punishments: they are given so that the dog will not repeat a particular behavior in the future. Situation and context determine which and when such "punishments" are appropriate. The rule we follow is always to use the least amount of force to change the behavior and firmly embed the lesson in the dog's mind.

Unfortunately, the word *punishment* has such a checkered pedigree that it is difficult to use in a clear, emotionally neutral way. Say "punishment" and most of us wince. We also tend to think of it as a consequence for bad behavior that can happen well after the fact. Criminals are punished for what they did in the past. In a relationship with a dog (a creature that lives almost entirely in the present), punishment after the fact is not only utterly unhelpful but destructive. For these reasons, we prefer to distinguish mild aversive actions from much stronger ones. For example, we use "corrections" to describe the light discipline that occurs in ordinary obedience training and everyday life. "Punishment" describes more forceful verbal and physical discipline associated with various behavioral problems. Not only is such terminology easier for people to accept, but we believe it is faithful to ordinary experience. Hopefully, the number of occasions you will need to take serious action with your dog will be very few, and only for major infractions. More typically, the discipline you rely on involves well-timed verbal and leash corrections that refocus and guide your dog, refining her understanding and reinforcing her perception of your leadership.

Notice that such corrections take place as an organic part of obedience training. Enlightened discipline is never divorced from the context of training. The problem with the approaches listed at the beginning of the chapter is that disciplinary action comes out of the blue; it is unrelated to a dedicated program of obedience training. We cannot count the number of times an owner has said something like "I'm not interested in his heeling or sitting, I just want him to stop growling," expecting us to provide a magic disciplinary technique. This is foolish. It is also ineffective — it doesn't solve the problem, and it hurts the relationship. Unless an owner is actively rearing a pup or older dog in a climate of positive training appropriate to the dog's age, any use of discipline tends to be imbalanced. Training is the way. The Royal Air Force Dog Training School has a saying that any aspiring trainer should keep well in mind: a handler always ends up with the dog he deserves. When divorced from training, corrections are harmful because they do not give the dog an understanding of what you want him to do. Discipline always needs to be followed by some sort of positive training command that reestablishes your leadership while also clarifying your intent to your dog. The traditional exercises themselves are a humane way of reinforcing your leadership and enhancing discipline. Putting a dog on a long down-stay, for example, not only gives you the control you may need at the moment but also reinforces in the dog's mind the fact that you are in charge.

## Preliminary Corrections

We discuss verbal and leash corrections in greater depth when we cover the obedience exercises later in this book. Here we will describe some general guidelines to serve as a foundation. Canines in the wild frequently vocalize to communicate with one another. They growl to maintain and exert authority. They also shoot assertive stares at one another. Often the alpha can stop unwanted behavior in the pack simply with a stern growl and piercing stare. The combination says, "Stop what you're doing . . . NOW!" It hardly comes as a surprise that what works so well for a wolf also works well for you. Ordinarily your eye contact communicates friendliness, trust, and love; the occasional piercing stare may well stop unwanted behavior dead in its tracks, even when your dog is not on leash, as does a sharp "no" or "nah." Circumstances govern the intensity and depth of tone you use, but the dog who recognizes your position of authority respects such corrections.

Any time your dog is on leash, a verbal correction should usually accompany a leash correction. Although some trainers recommend eliminating "no" from your vocabulary because it is so easy to overuse, we prefer to be more flexible. "No" or "nah" is a perfectly appropriate accompaniment to a leash correction, especially when followed up immediately with more positive direction and encouragement. Done right, this combination can serve to enhance and direct your dog's spirit and desire to work. Further, because "no" is entirely natural to us, the timing of the correction tends to be more precise: you don't have to think about it. And then there is a final point: though it is clearly different from training a dog, still, would you ever think of raising a child without the word *no*?

The leash correction, or "pop," is meant to refocus the dog when he is not paying attention. It is a correction used to communicate, not to injure, given to correct the dog's understanding of how he should be behaving at that particular moment. It is attention-getting and surprising, and is executed by giving the leash a quick pop and release, a gesture reminiscent of the motion used to snap a towel at someone at the local pool. The dog should find it mildly unpleasant, but it is not meant to be painful as much as surprising, to focus the dog's distracted attention back on you so that you can remind him what you are asking.

Besides the discipline that takes place in the context of basic obedience training, there may be occasions when more physical techniques are necessary to resolve specific behavioral problems. Our experience has shown that the following situations may merit both physical and verbal discipline:

Aggression with humans — defined as excessive barking, growling, charging, chasing, nipping, or biting a human

House soiling — defecation or urination in the house or in any other improper place

Stealing — theft of food or objects

Persistent destructive behavior — destructive chewing, digging, or house wrecking not the result of puppy antics or accidents

Aggression with other dogs — in-species fighting, usually between two males, but possibly between a male and a female or two females

To repeat, these situations *may* merit physical discipline. Since no book can pretend to analyze every individual situation, we feel obligated to repeat from the outset that physical discipline or correction is never an arbitrary training technique to be applied to each and every dog for all offenses. We do, however, believe that physical and verbal discipline can be an effective technique when used in conjunction with a broader program of obedience training. The safest policy if you experience serious manifestations of any of the above problems, or if you suspect your dog has serious behavioral problems, is to consult a qualified trainer or veterinarian to evaluate your individual situation (see chapter 8, "Where to Find Training").

If you decide upon discipline as a training technique, it should be the proper kind of discipline. No trainer can provide you with exact, surefire prescriptions for what correction to use and the amount of force needed. That has to be a personal judgment on your part based on knowledge and intuition. We can, however, provide guidelines. The following is our attempt, as responsible trainers, to map out several methods we have found both helpful and humane, methods that depend less on violent physical force than on timing, a flair for the dramatic, and the element of surprise. In considering their use, you should follow the rule of always using the least amount of force necessary to change the behavior. Don't go overboard. Build on your corrections, making them progressively tougher until your dog responds

*The wrong way to discipline a dog. Never use an object. Never discipline from above or behind.*

appropriately. Above all, watch your dog: his response will tell you whether the correction is too soft or too stern. Once you've obtained a consistent type of response, stick to that level.

As a prelude, keep in mind several general principles: never use your dog's name during a correction, never call a dog to you to discipline her, and never use an object of any kind to discipline your dog. Using your dog's name not only alters the timing of the correction but is apt to communicate something too personal, as if the dog herself, rather than her behavior, is what is objectionable. Calling a dog to you and then disciplining her compromises not only the recall but the relationship as well. The dog who associates coming with discipline not only won't come but will prefer not to be with you. As for using objects, many owners still rely on them when disciplining their dogs. The overwhelming choice seems to be the rolled-up newspaper. The hand that feeds is the hand that teaches and corrects. Do not use objects of any kind to discipline your dog. Just use your hand as described below. Finally, we repeat: physical discipline should be reserved for the heinous canine crimes mentioned earlier, not meted out for every episode of bad behavior. Verbal correction might suffice for many dogs, but you should know more than one method of discipline before the unfortunate necessity of using one arises.

## The Shakedown

One way to discipline your dog (and the first physical gesture to try) is the shakedown. It is particularly effective when you have been raising and training your dog (or pup) according to the principles of this book and suddenly are faced with a blatant case of testing or insubordination, such as snapping in a bratty way. The shakedown is a moderate physical correction that asserts your leadership and startles your dog into paying attention. The manner in which it is carried out depends on whether you are using it on a pup (grabbing the back of the neck, as his mother would) or older dog (grabbing both sides of neck fur).

In the shakedown the dog is sitting, anchored in place with tension on the training collar. When you have seated the dog and are sure he will not move, wheel around in front of him and kneel down. Grasp both sides of the dog's neck with both of your hands and lift him right off his front feet into the air. You may need to lean into the dog to do so. (When grasping a dog by his jowls, make sure that you have one or both of your thumbs looped under the training collar, to stop the dog

*The shakedown. Grasp both sides of the dog's neck fur as pictured and raise the dog's front slightly. Make eye contact and give a quick shake as you scold.*

from breaking away.) Holding the scruff firmly, look directly into the dog's face and shake the dog back and forth in quick, firm motions, gradually lowering the dog. Eye contact is essential. Scold the dog while you look at him, and keep him elevated a good five to ten seconds. It may be difficult to raise some larger breeds, in which case you have to sacrifice this part of the procedure. Most dogs, however, can be lifted up off their front feet with a little effort. Ideally, follow up immediately with a series of obedience commands to get the dog's mind back on your leadership. You want the dog to think, "Whoa, I definitely don't want to repeat *that!*"

After being disciplined in this fashion, the dog may be shaken up mentally and physically. Depending on the circumstances, you may wish to place him in an extended down-stay or simply heel without a lot of interaction. Keep your mood serious and businesslike until you have a chance to make up.

For young puppies, cut down on the intensity and duration of your correction. A young pup should be disciplined by simply grabbing the scruff of the neck with one hand and giving him one good shake. As we have pointed out, this method approximates the technique a mother uses to keep order in the litter, to stop fighting between litter members, or to help wean her pups away from her to solid food. Disciplinary methods that reflect instinctual canine behavior communicate displeasure in ways a dog can understand. Such corrections as throwing or hitting the dog with objects, spanking him with newspapers, or simple pleading serve only human, not canine, ends and do not communicate displeasure effectively to the dog.

## The Verbal Element in Discipline

Some dog owners find it difficult to say anything when disciplining their dogs, intent as they are on the physical manipulation required. Verbal scolding that accompanies the correction is essential. It requires a flair for the dramatic and good timing. The standard vocabulary for canine misbehavior is a growl-like "no, no," "shame, shame," and perhaps "bad dog!" Most of us have a whole speech prepared inside, ready to spill out, whenever we come upon or witness the results of our dog's misbehavior. We refrain from saying anything more than these pat phrases. Why?

One client put it this way: "Loddie does something wrong, and I know exactly what I want to say. I even know the tone of voice I want to use. But something tightens up inside of me." Another client said, "If I yell at the dog, the children think I am hurting her." Another dog owner reflected, "I've always been taught control over my temper. If you get mad, control it. Keep it all in. Although I am having tremendous problems with this dog, I can't see myself disciplining him physically or verbally. I just can't stand the feelings I get inside myself when I get angry."

The first thing we try to explain is that discipline need not be a terrible ordeal and that anger need not be a part of it. Dog discipline, if approached correctly and with a sense of humor, is more playacting than anything else, although the dog must not know it. On the other hand, the element of force is involved. In discipline, the owner puts the dog in a subordinate position and plays the alpha wolf in much the same way the leader of a wolf pack does. By keeping the vocabulary simple and natural, owners communicate more decisively and authoritatively. We encourage drama, timing, and surprise by having clients role-play situations when they might need to discipline their dogs. If an owner can successfully role-play discovering their dog peeing on the Oriental ten minutes before a dinner party, having a fight break out between his or her dog and the neighborhood rival, or any number of other catastrophic canine capers, then the actual occurrence of such events can be approached more naturally. So if you feel ill at ease with discipline, verbalize it ahead of time and role-play.

The tone of your voice is important and should be very sharp, intense, and commanding. Dogs are not deaf, so you need not yell or scream. Some dog owners are naturally verbally dominant, some are not, but a happy medium can be approached by all. Learn to say no, not in a whining, pleading tone, but as if you were throwing a verbal beanbag at the dog. If need be, go to a quiet place where you can be alone and practice belting out "no!"

## Physical Discipline Under the Chin

A second method of disciplining the dog, usually reserved for more serious offenses, is the cuff under her chin. This method is for older dogs and presumes a good relationship between owner and dog, in which leadership has been established. Upon the infraction the dog

*One way of disciplining an unruly dog is to sit the dog down and use an upward stroke under the chin.*

should be anchored in the sitting position. Your fingers meet the underside of the dog's mouth in an upward motion. It is essential first to sit the dog, by pulling up on the training collar or pushing down on the animal's rear end. This also rivets the dog's attention upward toward the owner's eyes, so that eye contact can be made. Eye contact is very important in discipline. Wolves disciplining each other make eye contact. Never hit a dog from above. Your fingers should be closed together, your hand open.

How hard do you hit the dog? A good general rule is that if you did not get a response, a yelp or other sign, after the first hit, it wasn't hard enough. One good correction will put an end to it. We have found this punishment particularly effective with various forms of aggressive behavior, such as when your dog growls at a stranger. The discipline is quick and decisive. A sharp smack under the chin followed by a quick string of obedience commands lets him know just how displeased you are with this behavior.

Keep one hand tightly on the training collar so that the dog remains sitting. Insert your index finger into the leash ring of the training collar, and wrap your fingers around the extension you will have when it is pulled snugly. Keep this tension as you discipline the dog with your other hand. Keep the dog sitting while you discipline and then quickly move into a series of obedience commands that reinforce your leadership.

To ensure that serious problems do not get out of hand, stage set-up situations later on in which the dog has the opportunity to overcome the specific problem. Having established a serious consequence for the inappropriate behavior (for example, growling at your guest), you will be in a position to positively reinforce acceptable responses through praise or treats.

## The Alpha-Wolf Rollover

In the original edition of this book, we recommended a technique we termed "the alpha-wolf rollover," to be used in conjunction with one of the disciplinary procedures already described. This disciplinary technique was nicknamed for the type of discipline the lead wolf dishes out to misbehaving members of the pack and involves following up an initial correction (shakedown or under the chin) with a down, then grabbing the scruff of the neck and sharply rolling the dog on her side and scolding her. The aim was to elicit a submissive response on the part of the dog that acknowledged your alpha status. This was a move we had seen have a powerful effect in certain extreme situations.

*We no longer recommend this technique* and strongly discourage its use to our clients. Though it can be argued that it has a natural basis in pack life, in a dog-human context it is potentially very dangerous and can set up the owner for a serious bite in the face (or worse), particularly with a dominant dog. The conditions in which it might be used effectively are simply too risky and demanding for the average dog owner; there are other ways of dealing with problem behavior that are much safer and, in the long run, just as effective.

Let us repeat: the disciplinary techniques explained in this section should not be applied haphazardly and for a slight misbehavior. There is always the chance that autocratic dog owners, having learned discipline techniques, will misuse them. Watch yourself — owners who are physically or verbally domineering wind up with cringing, neurotic dogs. Discipline, like praise, must be meaningful. It must communicate the owner's displeasure clearly, and on the dog's level of understanding and perception, for unacceptable behavior.

## Dogs Are Not Children

One woman came to us with a Pekingese with persistent chewing problems. She approached all dog problems as she would child problems. She quoted specialists who take the permissive approach to child raising. Some information on the dog's quite different background helped her see the differences between her Pekingese and a child. Obviously, there are some analogies between a dog and a child that can be drawn, but discipline is not one of them. Puppies in litters often discipline one another, jostling, shoving, and pinning down one another. The mother of a litter disciplines her puppies with gentle bites. Keep in mind that dogs mature much faster than humans, and their receptivity to discipline is in a different sphere from ours. Don't be afraid to discipline your dog, but do so correctly and on a level that allows the dog to understand the connection.

## Discipline Do's and Don't's

1. Go get the dog. Never call a dog to you and then discipline him. Even if a chase is involved, go get the dog.
2. Never use an object to discipline or punish.
3. Never use the dog's name in conjunction with discipline.
4. Discipline in proportion to the offense. Verbal discipline and leash corrections cover lesser offenses, while the shakedown and cuff under the chin are usually reserved for more serious, non-training-related offenses.
5. For the shakedown and cuff under the chin, first sit the dog. With the cuff under the chin, put tension on the training collar by inserting your index finger in one ring, and pulling it snug. If the dog doesn't sit, press down on her rear. Don't begin disciplining until the dog is sitting and anchored. Otherwise, the dog can more easily scoot away from you and avoid the correction.
6. Make eye contact with the dog as you discipline.
7. Never hit from above.
8. Remember to be dramatic.
9. Pay no attention to the dog for half an hour after disciplining.

## Making Up After Discipline

Just as it is important to administer meaningful discipline quickly and firmly when your dog commits a big offense, it is also essential to "make up" later. This making-up process does not have to be emotionally wrenching or cathartic. After you discipline your dog for a serious offense, remain passive for at least half an hour, not speaking to the dog and avoiding eye contact. If you attempt to make eye contact with your dog at this time, chances are that your dog will avoid looking at you. Your dog wants to make contact with you, but in his own time. Your dog behaves after discipline very much as a wolf does. Your dog may look away, gaze down at the floor, or look as if he were trying to melt into the wall. This is a natural reaction of submission that should not be interfered with by the distraught owner. Unfortunately, when the dog displays these reactions after discipline, many owners go to the dog and coddle him, trying to "cheer him up." When they approach, they may find that the animal remains motionless or trembles. This makes the owner feel even more guilty.

The fact is, the animal simply needs a certain amount of time to readjust. Just leave the dog alone for half an hour. On the other hand, if the dog shows no signs of submission or deference after discipline, you should reexamine how firm you were in disciplining. If your dog comes up to you and nudges you for attention five minutes after you have disciplined him, chances are that the dog didn't get the point.

After half an hour, do something friendly with your dog. Take a walk or a car ride, give the dog his favorite toy, or speak to him softly and encouragingly. Avoid dramatic make-up gestures, like food treats or robust play sessions. Some dogs may need more building up than others; you have to size up your own dog. The important point is to be big enough to make up but not so guilt-ridden and overbearing that you fail to give the dog the readjusting time he naturally needs after an effective correction.

# What's Cooking?

A sound mind in a sound body! Before we can go on to work on improving our pet's behavior, it is paramount that he be physically healthy. Really healthy, not just looking okay. Dogs can go on functioning well to all appearances when in fact they are in serious, poor condition. Later, just as with us, dogs, too, will begin to break down in their behavior, and many an owner will stand there concerned and baffled as to what has caused this change. The answer lies in the food dish.

In supplying our dog with a diet that maintains his condition at its best, we need a realistic view of the whole picture. Avoid taking economic shortcuts with generic dog food, as well as falling into sloppy feeding habits that can only hurt your dog in the final analysis. Feed the dog what he needs, not what you think he needs!

For most owners, a complete, nutritionally balanced dry kibble should be the basic dog food. It will contain correct proportions of protein, carbohydrates, fat, vitamins, and minerals. Particularly important is the percentage of protein as well as its source, which the label delineates. Because dogs are carnivores, our preference is for kibble whose main source of protein is meat. Many dry foods are soybean-based, and although soy is a rich and inexpensive source of protein, it is not easily digested by canines because it encapsulates other nutrients, thereby preventing them from being absorbed. Dog food companies are required by law to list the primary ingredient first, so check the label for meat. Many of the better brands are made from meat meal of chicken, beef, or lamb (and then secondarily grains) and may be available only through special distributors. These sophisticated feeds, though sometimes more costly, may be worth looking into since the companies that make them maintain research farms where nutritional experiments are conducted, giving greater confidence to their products. Also, higher protein content may mean less food (evening things out somewhat economically), resulting in smaller stool size. On the label, look for the seal of the American Association of Feed Control Officials (AAFCO), which guarantees content and proportion. Any kibble you feed a young adult of moderate activity should contain at least 22 percent protein, preferably more. Active dogs need higher levels of protein. Older dogs may need kibble with less protein content. Dogs who are overweight may require special feeds that are low in calories and high in fiber. As a general rule, watch for signs that may indicate your dog is not getting the diet he needs: excessively loose or large stools that smell particularly bad, bad breath, flatulence, low energy level, dull, smelly coat, and perpetual shedding.

Canine nutrition is a subject of much discussion, but overall we have had success with the finer kibbles and supplements. The reputable feed producer has usually included proper dosages of vitamins and other necessary ingredients in the kibble, but vitamins can often be lost in the manufacturing process through the heat involved in baking kibble, suggesting the need for some sort of supplementation. We supplement kibble with some eggs, cottage cheese, or a tablespoon of canned meat, plus some sort of commercial supplement that provides enzymes, beneficial bacteria, and vitamins (such as Ultimate Supplement).

Meat should not constitute more than 25 percent of the dog's diet. The American tendency is to go overboard on meat. Raw meat, frozen meat, tripe, canned meat, or table scraps — none should exceed more than one cup at each meal. Otherwise, you throw minerals out of balance.

Unrestricted feeding, the practice of letting the dog determine how much she eats and at what times, is problematic. Aside from the fact that such a feeding practice can lead to house-training difficulties, picky eating habits, and obesity, it also doesn't make sense from a behavioral point of view. By allowing your dog to feed herself, you take away the emotional bonding that arises from feeding your dog, which is an intimate daily contact that we believe is important. Instead, feed your dog at the same time every day. Two smaller meals are healthier than one large meal. A smaller portion is usually more easily digested. The risk of torsion (stomach bloat) in many breeds is serious and often fatal, and smaller portions help avoid this problem. Two feedings help prevent hunger tension, which can be a factor in problem behavior such as chewing, and can help cut down begging behavior between meals.

Puppies from the time of weaning to four months of age require three meals a day. The pup is growing in leaps and bounds, and his stomach cannot take an overload of food all at once. The feeding times need to be adhered to strictly and basically should be in the morning, at noon, and in the early evening. The evening repast should be no later than five or six o'clock, to give the puppy time to eliminate before retiring. Puppies generally need twice the number of calories and general nourishment of older dogs. Our experience is that they benefit from daily supplements of live-cultured yogurt (with acidophilus) or acidophilus powder, which helps restore intestinal flora; enzymatic supplements; and additional protein supplements of meat. A hard-boiled egg could be given several times a week in place of the meat.

A supply of fresh water is a must. Though a dog could go a number of days without food, without fresh water he would soon grow dehydrated and face serious health problems. The body weight of an adult dog is 50 to 60 percent water, more for a puppy. Every adult dog should have access to cool, clean water at frequent intervals, but keep water off-limits to dogs for at least thirty minutes before and after exercising. Puppies tend to gulp too much water and should be offered water periodically, rather than have free access to it. This is especially true during house training. If a pup or older dog spends time outdoors in the sun, he should have all the water he needs. Make sure that you clean the water dish or pail daily, as water left standing overnight can harbor bacteria. Water should not be allowed at night, especially if you are house-training a young puppy.

Never, ever refuse to feed your dog as a punishment. Behavior and health are related, but not to the degree that withholding food accomplishes any good in behavior. A dog has no consciousness whatever of the connection you intend by withholding food. All you invite is health problems. The feeding time should be marked by affection and praise. Since the dog's need for health is paramount, good or bad behavior should not have anything to do with regular meals. This may seem like simple common sense, but we have talked with some dog owners who try to prod their pets into good behavior by cutting off their basic rations.

## Treats

Treats fall into two categories. First, many people use food treats to reinforce behavior learned in the context of obedience training. This practice is fine as long as your dog does not become overly dependent on them (he won't comply without the prospect of a treat) or the food interferes with his ordinary feeding. We've seen trainers use pieces of hot dog or liver as rewards to the point where dogs have had the equivalent of a full meal over the course of a training session. Food treats do not represent a balanced diet and should not be used as a substitute for a meal. They are best used intermittently in training along with verbal and physical praise to reinforce and spur learning.

An occasional natural dog biscuit is entirely appropriate, serving both to clean the dog's teeth of tartar and to cement your relationship. You can also substitute a few pieces of washed, raw vegetables such as carrot sticks or green beans, which are not fattening and which dogs love to crunch. Stay away from soft, cookielike treats that contain chemicals and added sugars and food coloring. It's a good idea to ask the dog to sit before giving a treat. This takes only a second and helps the dog solidify the command. We strongly advise against treats at the dinner table; however, do not banish your dog from the room when eating. Teach the dog the down and down-stay first so he can be included in the experience. During meals your pet should lie off to the side or somewhere visible in the dining room. Keep the dog out from under the table and never let the pet sit at the table. It is not a "torture" for the dog to watch his owners eat — it is a pleasure to be present. If necessary, feed your dog first, before your dinner. A treat after dinner for a successful down-stay is fine, but whenever you treat the dog, indicate clearly when treat time is over. Tell the pet kindly but firmly that there is no more and to take his place.

Dogs, like people, can be fussy eaters. This can be frustrating since you are naturally concerned about the causes and implications of such fussiness. Don't lose your temper. Anger and pleading do not solve anything. The best approach is simply to set the dish before your dog for a few minutes and leave the dog alone. If there is no reaction, pick up the food and reoffer it later, at the next scheduled meal. Do not try to coax the dog to eat. Better to let the dog skip a few meals than to teach him to hold out for caviar. Normal, healthy dogs do not starve themselves, and it is better to get them to eat according to what you know to be best, at set times. In extreme cases, such as during illness or after anesthesia, hand feeding might work if you must get a few mouthfuls into your pet. But don't do so any longer than necessary, or your pet will come to rely on it. Always pay attention to your dog's basic demeanor when he doesn't eat, and try to notice whether something might be wrong with his health. Otherwise, a change in the weather or the addition of meat or a tablespoon of cat food (higher in protein than canned dog food) might perk up his appetite. If lackluster appetite persists and is not characteristic of your dog, see your vet.

How much food should your dog get? Begin by taking a good look at your friend. Is your dog lean or too fat? Don't rely completely on the statistics on the back of dog-food packages to gauge how much your pet should be eating. These set tables do not take into account the peculiar traits of many dogs, their high or low energy levels, the amount of exercise they receive, or their ability to metabolize certain types of food and should be taken as general guidelines. To determine whether your dog is too heavy, straddle your dog and put both hands over her ribcage. Prominent ribs indicate that your dog is underweight, whereas if they are difficult to feel, your dog may be too heavy. Try to keep your dog on the lean side. Extra weight means extra stress for the dog, who has to haul around those pounds. In pups, leanness and a pronounced stomach could mean worms, so it is wise to have a stool sample analyzed at your vet every three months until the dog is one year old, and every six months after that. Until you are comfortable with your dog's rations, evaluate him every two weeks to make sure he's not being over- or underfed. More books on canine nutrition are recommended in the reading list.

Though sometimes there are genetic or medical reasons for obesity, the most common one is the same as for humans: overweight dogs usually suffer from too much food and too little exercise. Dr. Nicholas Dodman reports that more than 25 percent of the dogs in the United

States are overweight, some of whom are obese (a body weight more than 25 percent above the ideal body weight).* Concern for our dogs' health and longevity demands that we look critically and seriously at our own role in canine obesity. In addition to traditional reasons such as the kids giving them food they don't want or your neighbors slipping them treats, our experience has been that many owners overfeed their dogs because of unconscious projections. Overfeeding is often a misguided attempt at compensating for something we perceive is lacking in the relationship (time, for example), an unhelpful way of making us feel as though we are really taking care of the dog. Overfeeding does your dog no favors: dogs who are fat die younger and have many more health problems. More exercise and less food means a healthier dog (and most likely a healthier you). In large breeds for which hip dysplasia is a frequent problem, too much weight can cause suffering and can further the progression of their dysplastic condition. Lastly, don't let the common myth that a neutered dog will gain weight deter you from altering one you don't intend to breed. Spaying a female or neutering a male should not increase the animal's weight, if the diet is properly controlled.

Commercial feeds may not do enough to help your dog's coat stay in top condition. Fats are required in the diet for good coat sheen. We have had good luck adding safflower oil or flaxseed oil to the food of dogs with dry coats (one teaspoon per meal), since these are the best source for linoleic acid, an essential fatty acid, and are the least allergenic of the common cooking oils.

Coprophagia, or stool eating, while repugnant to humans, is not uncommon in dogs and may indicate a problem related to digestion and absorption of food, dietary deficiency, overfeeding, or may simply be an obnoxious bad habit. For example, a dog being fed a rich diet once a day may produce stools with a high degree of undigested protein, causing the dog to snack later on. Because of the multiple possibilities, it is best to have a dog displaying this behavior examined by a veterinarian to make sure it is nothing serious. To deal with this problem, you need to supervise the dog's defecation times, and immediately after the dog has finished eliminating, call him in. Pick up after

---

*Dr. Nicholas Dodman, *Dogs Behaving Badly* (New York: Bantam Books, 1999), p. 50.

your dog immediately without your dog watching, especially if his area is enclosed. Make sure the dog is fed twice a day and that his stools are solid. Dogs who are confined to exercise pens and are unsupervised may be reacting to boredom or a lack of playthings. Enrich the environment with toys and other objects of interest to keep his mind active. Also, a number of folk remedies can sometimes help. Try sprinkling his food with Ac'cent. This addition seems to make the stool unattractive and may make up for the missing enzyme. Concoctions of Tabasco sauce and vinegar on the stools to deter your dog from munching have also been known to help.

## Problem Behavior and Nutrition

If you are experiencing behavioral problems with your dog, dietary changes might be beneficial. Contrary to long-standing assumptions of many trainers and veterinarians that low-protein diets can improve the behavior of aggressive and high-strung dogs, William E. Campbell has found just the opposite, provided that carbohydrates are reduced, too. In his nutritional studies involving problem dogs, he found that the general approach of feeding more protein and lowering carbohydrate intake has been effective in two ways.* First, conditioning to commands and signals seems to be better retained. Second, dogs appear less hyperactive and less disturbed by external stimuli, such as passing cars, other dogs barking, and loud noises. He suggests the supplement of a "high quality multi-vitamin/mineral/amino acid/ enzyme product" such as we recommend above, coupled with 100–500 mg of niacinamide to take care of any possible thiamin or niacin deficiencies that may be connected with poorly conditioned reflex formation and hyperactivity. He wisely calls such supplementation good "behavioral insurance." Our work with problem dogs bears out these findings. Vitamin/mineral/enzyme supplements make a difference with our dogs. Clients who use vitamins as a family and follow healthy eating habits themselves readily grasp this rationale; families with poor nutritional habits often feed their dogs poorly, too.

---

*William E. Campbell, *Behavior Problems in Dogs*, chapter 5, "Nutrition and Behavior" (Grants Pass, Ore.: BehavioRx Systems, 1999).

In cases of problem behavior, in addition to remedial training, we usually suggest suspending any "junk" foods that contain red dyes or high amounts of sugar. We suggest two feedings a day to prevent hunger tension. Finally, we teach clients how to decipher dog-food labels and the advertising psychology used by pet-food companies.

When you consult a veterinarian about nutritional problems or ask for advice, make sure you have written down exactly what you have fed your dog recently, no matter what it is. Don't expect a vet to provide you with helpful information unless he or she knows what you have been feeding your dog previously, what you can afford, and a basic health history of your dog.

# 12

# Grooming Your Dog

Some years ago we had a client who boarded her Newfoundland in our kennel whenever she traveled. She would plead with us also to groom the dog during his stay. We offered this service at the time, so we agreed, yet we can still recall how laborious and unpleasant a task it was. The dog was not used to being touched, his hair was tangled and matted, and he was always very smelly and dirty. After a few such sessions, we asked the woman why she didn't groom the dog regularly herself, since it would make her dog more pleasant to live with. She replied, "Oh heavens, no! Curly is just so big and unmanageable. I just can't handle him. If I try to groom him, he just whines and backs away. And his nails! Even our veterinarian has to anesthetize him to clip them. Curly just won't tolerate it otherwise."

This problem is common to many owners; it was allowed to develop because the owner didn't get Curly accustomed to being groomed as a young puppy. It is much easier to control a lively, 35-pound puppy than a reluctant 150-pound Newfoundland. The fact is, this principle applies to any breed. Regular grooming and the process of conditioning a puppy to enjoying it is an absolutely necessary part of caring for a dog, because grooming is critical to the overall health of a dog. It is not simply a cosmetic practice. Dogs whose nails are not regularly clipped can develop one or more of a host of problems, from nails growing back into the skin and difficulty in walking, to back problems in certain long-backed breeds. Dogs' ears that are not cleaned once a week can easily develop a buildup of wax, bacteria, and fungi, as well as full-blown ear infections. Dogs who are not brushed frequently usually begin to retain dead hair and often smell unpleasant; easily harbor fleas, ticks, and mites; and may be more susceptible to skin problems in general. Tear ducts that are not wiped regularly become encrusted with dried mucus (more commonly known as "gunk"), which usually leads to conjunctivitis. Dogs simply cannot take care of these things themselves: they need your regular attention. And just as with canine massage, regular grooming strengthens the bond between you and your dog while helping you keep track of what is going on with your dog's physical condition. As both you and your dog learn to relax and enjoy the process, you become more sensitive to subtle changes in your dog's condition, being better able to spot abnormalities such as tumors, skin or hair problems, and cuts or bruises. Given the expense of dealing with such conditions once they become full-blown, this simply makes good sense.

## Grooming Puppies

We start grooming our puppies regularly from very early on. For example, as soon as the second week it is important to trim puppy nails to make it easier for the mother to let her pups nurse. Generally at the same time we also go over the pup's body lightly with a slicker brush, just to start sensitizing her to this procedure. As the pups grow, we expand the session to include ear cleaning and opening the mouth (to examine teeth and palate). This is a part of responsible breeding that, aside from the health benefits, socializes and conditions the pup to new experiences. Had Curly been groomed regularly from an early age, his owner would not have had the problems with him as an adult and would have been in a much better position to enjoy his company.

If you purchase a pup from a breeder, ask how regularly the pup was groomed and inquire about any specific recommendations the breeder may have about grooming your particular breed. A long-haired, high-maintenance breed such as an Afghan, Old English sheepdog, briard, or a shih tzu requires daily attention to prevent matting, and you'll need several different and specific types of combs and brushes to keep the pup looking her best. Shorthaired breeds such as Doberman pinschers, vizslas, and many of the hounds usually require a weekly brushing with a soft bristle brush, while double-coated breeds such as German shepherds and huskies are notorious shedders who should be groomed at least twice a week with a slicker brush and shedding blade. At the height of the spring and fall shedding seasons, it is probably even necessary to groom daily. Ask your breeder: a good breeder is aware of the best way of dealing with your pup's grooming needs. If you have obtained a dog from a shelter, or an older dog from a nonbreeder, consider making an appointment with a professional groomer to show you the best way to care for your dog. This is particularly pertinent for owners of long-coated breeds.

## Establishing a Routine

We find that a regular grooming session establishes a pattern that your pup will come to enjoy and look forward to, which will carry over into adulthood. Begin with very brief sessions that are highly positive, fun, and relaxing. Talk quietly to your pup as you groom her. Follow up the grooming with a play session or a nice treat. Some owners may find no difficulty in grooming their pup on the floor; others may find bending over uncomfortable. If so, you may want to groom the dog on a nonskid surface such as a professional grooming table, a picnic table, or even a washing machine or dryer with a rubber mat over it. Think of safety: you don't want your pup falling from any kind of height. Keep a hand on your pup at all times and brush her lightly according to the needs of her particular breed. Be sure to brush all the way down to the skin. This removes dead hair and dirt and spreads the skin's natural oils throughout the coat and stimulates blood circulation under the skin. At the same time, check for fleas, ticks, scratches, and bumps.

A good way to determine whether your dog has fleas, aside from observing your dog scratching himself repeatedly (as well as flea bites on yourself!), is to run the comb over his coat with a white towel underneath. If dark flea residue shows up on the towel, treat your dog for fleas, even if you don't find any on his body. Though some people swear by natural methods of controlling fleas such as brewer's yeast and cedar shampoo, a more drastic approach is called for at certain times and places. We haven't had much luck with flea collars or powders, which don't seem to be effective at all. Systemic products such as Advantage and Frontline have been much more helpful in recent years. However, if you're suffering from a serious outbreak of fleas, we recommend combining one of these products with treating your dog's environment comprehensively. The life cycle of fleas from egg to larvae to adult is between three to six weeks, so to deal with the problem, you have to break the cycle. Adult fleas lay their eggs on your dog, and then the eggs fall off onto carpeting, upholstered furniture — wherever your dog rests. If you fail to treat (and re-treat) the environment, you're likely to miss more than 90 percent of the developing flea population, and you'll have problems again soon. If you really want to rid your home of the little pests, vacuum your house thoroughly every week, especially those areas where your dog rests as well as the edges of rooms and underneath furniture. Also wash or dry-clean all pet bedding weekly. If it should be necessary to use insecticides to control an infested house, check with your veterinarian to make sure that there is no conflict between products used in the environment and those used on the dog or systemically.

Ticks are another parasite you must check for regularly, particularly if you take walks with your pooch through wooded or grassy areas. Should you find a tick on your dog, use a tick remover to pull it out. This device grasps the tick at its head and, when twisted, removes the tick completely. Deposit the tick in a jar of rubbing alcohol. It is important to pull out the whole tick, body and head, since getting just the body leaves a higher risk of disease and infection. Deer ticks, which are very small, carry Lyme disease, which causes lameness and intense joint discomfort for humans and dogs. If it seems like a deer tick, save it in alcohol and have your veterinarian check it for Lyme disease.

After you've groomed the dog's entire coat (including her legs, feet, and tail), next check the ears and eyes. Gently massage her ears and then inspect them for dirt, scratches, redness, and parasites. Be alert to any sort of foul odor coming from the ears, which could signal an ear infection. To avoid digging down into the ear canal (which can be harmful to the dog), we spray an ear-cleansing solution into the ear (a fifty-fifty solution of apple cider vinegar and warm water works as well), let the dog shake, and then daub the area with a cotton ball. Never probe farther into the ear than you can see. Cleaning should be done weekly. Next, check the eyes for redness or other signs of irritation. The eyes should be clear and the eyeball white. Wipe away any mucus discharge from the corners with a cotton ball moistened with warm water, and praise your dog as he allows you to do this. Make sure not to wipe the cotton ball over the eye itself, which can irritate it.

An important part of a grooming routine is checking your dog's mouth. Inevitably, at some point in his life, your dog will need to be given medicine orally. Doing so is much easier when the dog is accustomed to having his mouth opened. Enfold his muzzle with both of your hands, one on top and one underneath, and open his mouth gently, taking a quick peek at his teeth. Be very appreciative and full of praise as he allows this. Though it may seem excessive to some, many veterinarians are also now recommending regular teeth-brushing sessions to keep plaque under control and preserve good dental hygiene. It also controls "dog breath." Should you choose to brush your dog's teeth, break your pup in very slowly and use a toothpaste intended for dogs. Special brushes or small pads are available to make it easier for you, and more effective.

Last, but definitely not least, is trimming your dog's nails. Believe it or not, this doesn't have to be a perpetual battle of wills. If you handle your pup's paws and toes briefly every day, right from the start, he will become quite relaxed about having his feet handled, especially when you combine it with something he finds enjoyable (such as a belly rub). The secret is being consistent from the start. Since puppy nails grow quickly, it pays to clip just the very tips, but do so often. Use a nail clipper specifically designed for dogs, and be sure to trim just the tips. This avoids cutting into the quick, which is painful to your dog and causes bleeding. If you are dealing with an older dog, you can desensitize him to nail clipping much more gradually. Sit on the floor with him and gently take his paw in your hand and lightly massage the toes. When he allows you to do this without pulling his paw back, praise him warmly and give him a treat. Practice this for several consecutive days. Next, begin applying light pressure to a nail by tapping it lightly with your index finger. As the dog allows you to do this, give him a treat. Be prepared to go slow — there's no rush in view of the long-term goal. Once your dog is comfortable having his nails touched, show him the nail clipper, letting him sniff it if he is inquisitive. Gently hold his front paw with your free hand and clip only the very tip of one nail. Praise him warmly and give him a treat. If your dog seems comfortable, trim the remaining nails on the paw in the same way, rewarding him after each clip. Conclude the session once the initial foot is finished. Do the remaining paws over the next several days in a similar fashion. Usually this approach can condition a dog to the nail-trimming procedure in a short while, then all of the nails can be done in one session. The key is to remember always to praise and treat, and to take off just the tips. This prevents painful bleeders that cause dogs to react negatively to nail clipping. (Powdered alum, liquid stiptic, or a stiptic pencil are available at pet stores in the event you accidentally cause a bleeder.)

## The Older Dog

Over the years in our work with dogs, there have been times when we've imported an adult dog from Germany for our breeding program. Usually it has been a very easy process of adaptation into our program, but occasionally we've discovered that the dog we've imported has not been trained/conditioned properly to regular grooming. Initially, she may be uncomfortable being touched and handled. The same may often be the case when someone obtains an older dog. We can apply the same principles to condition an older dog to the grooming process that we do for conditioning an older dog to nail clipping.

We first try to determine the level of sensitivity to touch. With many dogs the problem isn't so much being touched at all as it is being touched in particular areas, such as the paws, sides, rump, or ears. With the dog on leash, start by petting him down the back of the neck, then gradually test other areas of his body. Once you determine those areas that the dog does not mind being touched, focus your attention there. As the dog relaxes, gradually graze the area he is sensitive about, going back immediately to the area of comfort. Praise and treat. Keep these initial sessions brief. It is better to have a short session five consecutive days than one long one. Always remember to praise warmly after a brief brush in the area of discomfort. As the dog signals acceptance of your touch, extend the number of touches. Assuming that you proceed with patience and sensitivity, the dog should become accustomed to being touched in sensitive areas fairly quickly. This will then allow you to begin formal grooming.

However, first make sure that your dog is free of significant mats and tangles. There is no question that a major reason older dogs dislike grooming is that their initial experiences with it have involved pain. If your dog is seriously matted, have her professionally groomed first, then work on getting her used to grooming gradually. Begin with light brushing in small sections (opposite direction first), taking frequent breaks and providing plenty of praise and a treat. Do not be concerned with grooming the entire dog in one session. Dogs can get very fidgety and then become hard to control. Better to do it over the course of a couple of days and preserve a pleasant attitude. If at any time your dog begins to put her mouth on your hand, give her a quick pop on the leash and say sharply, "Ahn!" Then go back to brushing, praising her warmly as she permits this. If your dog is a struggler, get a family member or friend to gently assist you for the first sessions, and keep them very short. Once your dog has been through a series of successful brushings, you can try to extend the length of the session (as well as groom solo).

## What About Baths?

Years ago, when shampoos were harsh and robbed the dog's skin of vital oils, dog professionals believed that dogs should be bathed minimally, only when dirt and filth made it absolutely necessary. Now, with milder shampoos that keep the pH level of the skin balanced, dogs can be bathed as often as once a week without drying out their coats. Generally paying attention to how your dog smells may very well determine bath frequency. If there is a "doggy" smell, time for a bath!

It helps a lot if you brush your dog thoroughly before the bath, to remove dead hair and mats. Doing so prevents a mess of tangles as you wet the dog. Some people use either a bathtub or walk-in shower (where it may be easier to keep the dog confined); others bathe their dogs outside. You need shampoo, towels, cotton balls (to put in the ears so water won't flow into them) — all within easy reach from wherever you intend to bathe the dog. If you have a movable shower attachment, use it to bathe the dog. First wet down the dog's entire body with lukewarm water from the neck down, keeping one hand on his collar to keep him from bolting. Some soothing and encouraging words help keep things calm. Work the shampoo into his hair thoroughly, then rinse and repeat the process. Next, when you wash his head, make sure the cotton is inserted into his ears, and be careful not to get soap in your dog's eyes. By doing the head last, you help keep the dog from shaking the water out and keep yourself drier. Wet down the whole head and work a little bit of shampoo on the top of his head, behind the ears, and around the muzzle. Rinse gently but thoroughly. As we finish, we like to work as much of the residual soap and water out of the coat by running our hands through it, then hold the towel up along the side of the dog (so that it protects you) and let him shake several times. Towel the entire dog — head, coat, feet, and tail. Either use an electric hair dryer made for dogs or make sure you keep your dog in a warm place (in colder weather) until he is completely dry.

# 13

# The Inner Dimension of Training: In the Beginning Is the Relationship

Obedience trainers can be awfully demanding tutors (some are notoriously blunt about pointing out their clients' mistakes), yet they also delight in those occasions when they see their work with a client become inspiring. To observe the relationship between dog and owner blossom is one of the most satisfying things a trainer can experience. It makes all the routine work well worth it. One time a client brought her golden retriever for our three-week basic obedience-training course, even though she and her dog, Jessie, were not having any serious problems in their day-to-day relationship. The woman explained she felt the relationship could go much deeper, and she was willing to spend the time and effort necessary to achieve that goal.

We suggested several books and articles to help her explore further possibilities while her dog was with us being trained. We also assured her that we would delve into many of these points more thoroughly at the final interview. As it happened, when she came back for that final interview she was well prepared; she had done her homework and had thought clearly about how she wanted to improve the relationship. She was enthusiastic, not only about the technical routines and insights of the obedience course but also about our comments on the inner discipline required of owners if they are really going to move to the next level with their dogs. During the demonstration itself, she paid close attention to the monk working with Jessie, asked her boyfriend to videotape the session, and was obviously overjoyed when finally she was reunited with Jessie. After spending some time playing and simply being with Jessie and as she was preparing to go over the training along with the monk, she paused to make an observation. "You know, Brother, I've been studying ballet for a long time now, and the way you were working with Jessie reminded me of things I know intuitively from dance. It seemed as though you were working effortlessly, as a team. And her attitude! I was so relieved. My friends had warned me about training, but far from breaking her spirit, she's really come out of herself. With this foundation, I feel I've got something tangible to build on and a certain quality of relationship to shoot for."

Her comment was gratifying, and not simply because it was complimentary. Far more significant was that she had keyed in to a deeper dimension possible in her relationship with Jessie, one with clearly spiritual overtones, and was willing to work to develop it. Her image of ballet captured well the harmonious ideal of the canine-human interaction: disciplined yet graceful, balanced and always respectful. It was a real pleasure several months later to receive a card from her saying that although "not quite choreographed, our life together is so much richer! Thank you for your help."

Our approach to dogs at New Skete places the main emphasis precisely on fostering the relationship between dog and owner. As we begin to teach the basic obedience exercises, it is important to underscore this principle: training is always intended to serve that relationship. A healthy relationship comes from a thoughtful and respectful stance toward the dog, in which we as caretakers combine solid understanding of dogs and dog behavior with our deeply felt love for them. Yes, good technique is important, but not for its own sake. It always takes a secondary role, supporting the overall quality of the relationship. In our experience, such an attitude on the part of the caretaker frees the dog to respond with a natural willingness and over time nurtures a profound level of loyalty and dedication, despite the fact that there are many valid methods of training a dog. This is the fertile soil out of which so many famous dogs (Lassie, Benji, Rin-Tin-Tin, etc.) have developed: a sound relationship with the trainer.

This comes as a surprise to many people. There is so much emphasis on technique within our culture that we habitually presume that if only we can get our position correct, the timing of the leash pop precise, the tone of our voice at such a pitch . . . everything else will follow. There is more to training than that. Our attitude is equally important in effective training, in communicating and building the relationship you envision with your dog. Training is a challenge because it involves two living, conscious parties: you and your dog. In our experience, most dogs act fairly predictably, according to their personalities. However, we human beings are often less predictable. It is all too easy to be unconscious of yourself and bring into the training process many kinds of contentious, negative attitudes that are counterproductive, that dump on the dog, and that ultimately hinder any progress. This happens, for example, when the dog receives the aftershock of the fight you had with your spouse, the reprimand you received from your boss, or simply the bad mood you happen to be in that day. Your dog has no way of understanding these attitudes beyond the immediate message your behavior is communicating: you are displeased with him. Not only is it unfair to you and your dog, it is not even really true.

The good trainer has to be so aware of both him- or herself and the dog that he or she can respond immediately to whatever the circumstances may require, cleanly, without any unrelated emotional "baggage." To do so requires a ruthless honesty and disciplined introspection that make you aware of your state of mind as you prepare to train. This awareness helps you identify and deal with your own personal emotions in advance, which can affect your training. Far from being artificial, amazing things begin to happen once you bring these healthy characteristics to your training. Not only do you have a clearer focus and better presence of mind toward your dog (which helps the dog learn in a much more efficient manner), you also carry these qualities over into your human relationships as well. The two spheres are intimately related. We have a saying here at the monastery: you can't become a better trainer without becoming a better human being.

From this perspective, obedience training can be allied with spiritual discipline both to train your dog and to train yourself. The difficulty with many approaches to training is that they focus entirely on the dog and not on the trainer. In such a context of unreflective behavior, all that really matters is the particular technique and the end result it achieves. The trouble is, such overkill has a profound effect on the relationship. "Whatever works" is an excuse used to justify training techniques that, when looked at closely, not only lack sensitivity but are ethically questionable. For example, "hanging" a dog for an obedience infraction may seem to get results, but at what price? How does the dog now perceive you? As a tyrant worthy of fear? What other issues in your life are coming out in your anger and frustration toward your dog? Is it really worth it?

We believe there is a better way, one that combines a genuine respect for the dog and how he learns with good technique and a consciousness of the actual interior state of the trainer. It is important to realize that this inner dimension can be worked on and acquired by most of us. We are not fans of "the Saint Francis syndrome," the notion that some people simply have a way with dogs while others, no matter how hard they try, will never be able to learn. Though it is true that some people intuitively sense what is helpful, it is always possible with effort and reflection to improve one's attitude and training skills.

This is not a book on religion, and we have no desire here to go the unwise route of mixing religion with animal care. At the same time, we would be remiss if we did not observe and clearly reveal that there is a deep spiritual connection between dogs and human beings, which can be enhanced and developed when your relationship with your dog is grounded in some form of personal, spiritual discipline. Meditation is valuable in any human life, because it teaches us how to be calm and fully conscious of what we are thinking and feeling. This experience and personal knowledge enables us to be more appropriate and effective in our communication and behavior. This is of crucial importance in dealing with a dog.

For example, dogs are highly sensitive to human emotions and are equally perceptive in reading our body language. It is difficult to disguise a bad attitude with a dog. Attitude comes through loud and clear and affects the dog. No matter how good our technical training skills, if we are angry, whiny, impatient, lacking in creativity and spirit, the dog perceives and reacts to it, and our training suffers. To avoid this, it is of extreme importance to create a genuinely positive, encouraging attitude in working with your dog.

First sit down and be quiet for a few seconds. Focus and center yourself by taking a few deep breaths. Let go of any tensions and anxieties you may be harboring and move your attention onto your dog. Become aware of the actual concerns you have for your dog and let that be the foundation from which your training proceeds. When you feel clear, calm, and relaxed, try to visualize the session you intend to have. Recall what you had difficulty with in the previous session and then go through the exercises one by one in your imagination, trying to anticipate possible scenarios that may take place, as well as the response you would hope to provide in each variation.

This does two things. First, it directs all your inner energies toward the session, frees you from distraction, focuses your attention, and makes you alert and ready to work with your dog. Second, it helps you familiarize yourself with exactly what you are going to do. Training is dramatically more effective when it flows naturally, when the timing of corrections is precise and you don't have to stumble over your own thoughts and actions. Many athletes use exactly this same technique before competition. Mentally going over what they hope to do in the coming event prepares their bodies to act naturally.

Think positively, and the dog will respond positively to your overall demeanor, inner and outer. In the past it was assumed that dogs respond obediently in a training situation but not in real life simply because they recognize the setup. By removing the possibility of noncompliance (for example, using a rope with a long-distance recall), the dog understands that resistance is futile and thus will come. However, another hitherto unacknowledged aspect enters in here: the handler's relaxed attitude during the setups. Because they are not worried that the dog won't obey, handlers usually relax in a way that isn't the case when the dog is free. The dog immediately picks up on this and is more likely to respond obediently to the handler's confidence. This is a crucial dimension that realizes its fulfillment once the dog is fully trained and the owner confident that the dog will be obedient: the relationship naturally manifests that sense of trust and harmony.

A truly positive approach to training is holistic in character, conscious of and addressing the psychological, spiritual, and technical dimensions present in the dog-human relationship. Paying attention to these elements together is what moves the relationship to the next level. The owner assumes a position of informed responsibility for the relationship and views training from the perspective of communication, how best to help the dog understand.

A good attitude does not mean that there won't be limits set or consequences for undesirable behavior. Just as with human relationships, human-canine relationships, by their very nature, are not one-dimensional. They involve the healthy and constructive use of correction and guidance, practice and play, companionship and periods of being alone. Being aware of all these dimensions is the first step to integrating them into the relationship. These elements mirror pack relationships in the wild, and their cumulative effect is to accent the positive, to motivate by encouragement and patience.

However, such relationships are founded on healthy respect. As we have learned in our own monastic life, true obedience involves mutuality and reciprocity. It is not simply the dog that obeys. The root meaning of the word *obedience* is "to listen." When applied to training our dogs, obedience involves as much our listening to the dog in order to discern what is needed as it does the dog's responding to our commands. It involves laying aside our burdens for the moment and entering fully into the relationship here and now so that our word to the dog will be simple, clear, and free of emotional or physical static.

# Environments

# 14

# Canine Environments

There are four basic types of canine environments: the city, the suburbs, the country, and outdoors. Each presents a different lifestyle for the dog, and each holds the potential for problems. Contrary to popular belief, there is no ideal setting for each and every dog. The myth that no dog can live happily in a large city is being debunked daily as thousands of dogs manage to coexist with smog, skyscrapers, noise, pollution, and the lack of space. The reverse myth that a dog can find true happiness on a farm certainly doesn't apply to all dogs, since country dogs are presented with a different set of problems.

We do not pretend to point out every pitfall in each particular setting, but we do want to identify some of the main problems specific to each. Some dogs spend part of their lives in all four settings, and others move dramatically between city and country, even in the course of a year. No matter where they live, smart dog owners avoid the temptation to blame the locale for their dog's problems and behavioral quirks. Though different environments may trigger idiosyncrasies, training can usually help your dog overcome them.

Don't be afraid to have a dog simply because of where you live. Dogs can be happy almost anywhere if their lives are properly structured and if they are conditioned to the particular environment and its demands. Finally, owners must take a hard look at their canine's environment and try to make it as healthy as possible.

# 15

# City Life

Is it impossible to own a dog in a large American city? Apparently, the answer is no, as thousands do just that. They own dogs of all sizes and breeds, not just the typical apartment dogs. As the crime rate increases in some cities, many people purchase large dogs with a "protection image." German shepherd dogs, rottweilers, and Doberman pinschers have become popular in urban areas, and poodles and other smaller breeds have been living there for years. The fact is, some of the best-cared-for dogs attending our training program over the years have been owned by clients who lived in large cities such as New York or Boston. Because of the challenges involved, they have had to be utterly serious and realistic about the commitment required to care for a dog in the city. Their example convinces us that it is possible to have a happy dog in a large city — but it takes time, dedication, and money.

One of the most obvious problems is providing your dog with a balanced amount of exercise every day. If you are in the city and own a dog, you need to commit yourself to a program of two, and preferably three, exercise outings each day. For owners who work, midday walks by professional dog walkers make this possible and have a marked influence on the dog. The length of the walk depends on the breed. These walks should entail more than simply time outdoors for the dog to eliminate. They must be exercise outings in which the dog is walked at length or allowed to run. The walk itself should be taken on leash, since almost all cities have strict leash laws. Your dog needs to go out every day, fair weather or foul, summer and winter. If you live in an upper-floor apartment, there is no convenient way for your dog to go out and come back in on his own: the animal's access to the outdoors depends on someone. There's no way to avoid the responsibility of taking the pet out.

The exercise problem is one of the most obvious challenges to keeping a dog in the city, and yet it's not all that bad. No matter where a dog lives, he needs the same amount of exercise and needs to go out to eliminate. The major difference in the city is that these activities must be regulated on leash, both when you are home and when you are at work. Many people find it helpful to set up a daily schedule fitting in some kind of exercise for themselves along with their dogs. In New York, for instance, joggers hold their dogs on leash as both run around in parks. In some large cities, groups of dog owners have cooperated in installing large dog runs or dog parks where dogs can be brought for exercise, play, and elimination purposes while the owners read or chat. For a dog with the right temperament, these are wonderful areas where they can romp and play with other dogs off leash. Make sure, however, that the run is well supervised and that the dogs playing are good with other dogs before allowing your own to romp. While he is playing, keep an eye on your dog so that he, too, stays within acceptable bounds with dogs and people. For those with large or extremely active breeds, a program of roadwork (see chapter 26, "Keeping Fit for Life") provides the dog with exercise even if the owner cannot walk along. But in the city roadwork is most effectively accomplished with a springer attachment. This piece of equipment attaches to your bike and allows your dog to trot alongside on a nylon line. Such roadwork can be done safely and easily in a city park where traffic is minimal. Lastly, if you Rollerblade well, you can also teach your dog to move at your pace, whether on heel or out in front in a controlled manner. Simply understand that with the latter two exercises, it may take a little bit of time for your dog to adjust, so take it easy.

One of the most important skills you can teach a city dog is to fetch. Ten minutes of fetching can provide a hearty exercise session for most dogs and utilizes many more muscles than merely walking. A fetch session should be on the daily agenda of every city dog; however, make sure to use a long rope or nylon line that can trail behind your dog as she runs after her "prey." This ensures that you are always in a position to enforce your dog's recall should she start playing "catch me" games with you. Bring the rope with you in a canvas shoulder bag along with something to fetch. Frisbees, sticks, or a deflated football are objects that can be located more easily than a regular ball after a throw. Every big-city park must be strewn with an assortment of balls intended to be retrieved by dogs but eventually thrown too far and lost. Young people discovered Frisbees in the sixties, and they are extremely popular with dogs now; in some places Frisbee catching has become an organized competitive event.

# Professional Dog Walking in the City

Increasingly seen in large cities these days are groups of dogs being walked by a dog walker (occasionally with an assistant). Such "pack walking" can involve as many as ten to fifteen dogs, and although it looks pretty impressive, people should understand its dangers. While we vigorously support the idea of dog walking, we recommend that the number of dogs being walked at any one time not exceed four, unless a larger number are from the same household and have been trained properly. The only bonus in walking a greater number of dogs is to the dog walker, in the form of more profit. The risks are far more serious:

- The more dogs being walked, the greater the possibility of your dog getting loose. In such an event, the walker would be unable to chase the runaway because of the other dogs he is walking.
- Pack dynamics can lead to aggressive behavior. The more dogs, the greater the chance for aggression both within the pack and toward another dog the pack happens to come upon.
- Dogs are often tied up alone on the sidewalk when the walker is fetching the next dog to add to the pack. Dogs that are being walked should not be left unattended.
- Pack walking usually involves dogs of different sizes. Dogs that are smaller or younger may have trouble keeping up with the larger ones. This can lead to overexertion and excessive stress.

These caveats aside, a reliable pet walker or pet sitter is invaluable to maintaining a healthy dog in the city. With such a professional you can leave your pet in her own secure, familiar space when you are away, either during the workday or for longer periods, such as vacations. Like people, pets are creatures of habit, and by staying in their own home, they are able to follow their normal eating, medication, and exercise routines. Such familiarity contributes to the happiness and health of your pet. There are other advantages as well. When using a pet sitter, for example, you minimize the possibility of exposing your pet to illnesses she may come in contact with at a kennel.

Use common sense in choosing such a pet-care person. Solicit referrals from reliable sources, such as your veterinarian, local animal shelter, and neighbors who may be acquainted with (or use) such individuals. Set up an interview and ask the person to provide some references. If the interview involves dog walking, ask how many dogs will be walked at once and, if you are hiring through a service, find out whether the service is bonded and insured (a nice bonus if it is). During the interviews, pay attention to the prospective walkers' general demeanor, the questions they ask, and the way they interact with your pet. Watch how your pet responds to them. Do they ask specific questions about your pet, or do they seem more intent on simply selling themselves? Be clear about the projected length and responsibilities of their work, as well as the costs involved. Dog walking is for at least half an hour, more probably for an hour. Inquire about how they would handle an emergency, what steps they would take, and whom they would contact. Ask if they are open to beginning with a trial period, which would allow you to get an idea of how your pet is handling the service. Finally, take a walk together with your dog to observe how the walker handles your dog. This will help your pet relax and adjust to the dog walker.

If you agree to go ahead with the service, have your keys stamped DO NOT DUPLICATE and make sure not to attach your name and address on the keys. It is better for dog walkers to color-code keys to protect the owner's security should the keys ever be lost or stolen. Lastly, have copies of emergency telephone numbers (veterinarian, your work number, neighbor, etc.) near the front door or on your refrigerator.

"I wouldn't own a dog in the city — he would never get any fresh air!" This common complaint has the same merit as that which applies to human beings. Air pollution in some cities is unremitting, and the air inside is usually cleaner than the air outside. Take the normal precautions for your dog that you take for yourself. Don't expose the dog to noxious vehicular exhaust if you can help it. Try to get your dog out onto grass at least once a day, even if it is only a small patch of greenery. Never blow cigarette smoke in your dog's face, even in jest. Also, many city dwellers (e.g., New Yorkers) take off to small homes in the country for weekends, thereby providing their dogs with environmental balance.

Noise pollution is actually more of a problem to dogs than air pollution is. Trucks, honking horns, sirens, airplanes, sonic booms, and crowd noise all take their toll on the city dog who has not been sensitized to various noises. Use the same precautions for your dog as you take for yourself. Avoid walking past construction projects if you can, because loud, sudden machinery noises can make even the best-trained dog break heel and dart away. Narrow, thin alleys or precarious, temporary construction crossings can be a problem.

Brian Kilcommons makes an interesting suggestion that is applicable to young pups and fearful or hesitant dogs: play sound tapes (e.g., of thunderstorms or loud city sounds) in your apartment while your dog is eating.* By starting at a very low level, you can gradually accustom a dog to tolerate such sounds without concern. Remember, not all dogs are of the caliber of Seeing Eye dogs, who are able to cope with city stress because they are genetically and educationally prepared. When you must traverse an area where noise is deafening, hold the dog near you on leash and cup one hand around the dog's neck until the noise dies down or you pass out of its range. This comforting body contact can help the dog cope with the noise more easily.

Dogs should not be allowed to run free in city parks unless they are completely controllable. In some cities the law may forbid off-lead dogs altogether. Even city parks can be a stress for a dog and can provoke strange behavioral reactions. For instance, never allow a dog to run free with strange children. Don't allow noisy children to crowd around your dog. Groups of screaming children have often triggered biting incidents or encouraged playful dogs to jump up. A child who is jumped on in play and then falls and screams can be perceived as prey by a dog, with occasionally tragic results.

The city dog needs to be able to deal with an incredible array of strangers each day. Many humans in cities simply go on "automatic

*Brian Kilcommons with Sarah Wilson, *Good Owners, Great Dogs* (New York: Warner Books, 1992), p. 220.

pilot" and pass strangers without seeing particular faces. This ability to screen out distractions is more difficult for dogs to acquire, especially with their highly developed sense of smell. The dog remains interested, in a positive or negative way, in practically every human and dog he passes. Pedestrians may react in a variety of ways, from fear to over-effusive affection to outright disdain or hostility. There is simply no way to predict their reactions, so the best approach is to expose your dog to all possibilities in a structured training session. Accustom your dog to being approached, petted, and also, possibly, rebuffed.

A leadership role by the master, and heeling practice, can help rivet the dog to his owner, but be aware that heeling is always more difficult in the city. There are simply more distractions, more opportunities to lag behind and investigate or to lunge ahead. If you follow the heeling methods described in this book and train your dog to heel by using distractions in your training sessions (traffic, other dogs, crowds), your dog should be confident and controllable and able to handle any situation on the street.

In an elevator, you and your dog might be squeezed in with a crowd. Accustom your dog to riding in an empty one first, before attempting to ride in a full elevator. If your dog either is prone to aggressive reactions or simply wilts if caught in a crush of people, you have to pay extra attention in elevators, crowded hallways, and rush-hour crowds on city streets. If this is a problem for you, keep your dog on leash and his training collar high around his neck for more control. Gradual exposure to these situations often improves a dog's performance.

No need to mention the foolishness of allowing a dog to run free in an urban area, or of ever walking the dog off lead, for that matter. Both practices are extremely dangerous for city dogs. A free-roving city dog can galavant around continually, chased by dog catchers and others. Life for this dog soon becomes the equivalent of guerrilla warfare, a daily ritual of scavenging, fighting, and avoiding capture. Studies on the behavior of free-roving urban dogs have shown that these dogs learn to move quickly.* They can be so cunning that they avoid capture for weeks. If you live in a congested area and let your dog run free,

---

*See Alan M. Beck, "The Ecology of 'Feral' and Free-Roving Dogs in Baltimore," in *The Wild Canid: Their Systematics, Behavioral Ecology and Evolution,* ed. Michael W. Fox (New York: Van Nostrand Reinhold, 1975). Mr. Beck estimates that "there is one free-roving dog for every nine humans in Baltimore."

unsupervised, he may be recruited into one of these canine gangs. The end that awaits these vagabonds is the pound and probable euthanasia. As for walking your dog off leash, there are simply too many dangers in a city ever to make doing so advisable. In New York we noticed a gentleman walking his Doberman pinscher off lead down the sidewalk. The swagger with which he walked indicated how proud he was of his dog, and the dog was clearly well trained. We could only shake our heads at the man's foolishness. Dogs are not perfect. All it takes is one mistake — a cat or other animal running out suddenly and the dog chasing it out into the street — and the result could be catastrophic. There is no shame in walking your dog on leash.

Since city dogs are so restricted, they often wind up staying at home alone. It's just not possible to take the dog everywhere in the city. Even if the dog can stay in the car comfortably (though it's often too hot for her to stay in the car alone safely), in some cities there is a very real chance that the dog could be stolen. On the other hand, most city pets serve a need for protection and security, so they are left behind to guard the owner's belongings. The resulting isolation, accompanied by a backdrop of urban noise, is often a prelude to incessant barking, destructive chewing, or other frustration-release activities. The situation spirals downward as continual barking or whining leads to complaints and possibly eviction. Destructive chewing can cause an apartment dweller to forfeit an expensive damage deposit, not to mention the loss of personal belongings.

If you live in an apartment, carefully consider which rooms the dog will have access to when you are gone. Some owners may need to train their dogs to eliminate initially on newspapers in the kitchen, then wean the dog to eliminating outside as he gets older. During this time, we advise confining the young dog to the kitchen by using an expanding gate, then crating him until he is fully housetrained. If the dog is being seriously exercised at midday, this is not at all unreasonable. Generally, we do not expect a dog to be capable of having full access to the house until he is between a year and a half and two years of age.

In apartment buildings, most bathroom and kitchen pipes and ventilation systems connect to upper and lower floors. A dog that barks in the kitchen, bangs her tail on the bathroom radiator, or yodels in the living room will probably disturb several residents of an apartment complex. The terrace is no place for an unsupervised dog. Some city owners train their dogs to eliminate on a terrace, sending fumes, aromas, and even droppings down or across to their neighbors. Others use the terrace for exercise, and run the risk of the dog's falling or even hurling himself over the railing.

Correcting these problems can be difficult, regardless of the techniques used, since in some cases the dog simply cannot tolerate the city environment and meet its demands. Obedience training, at least to the heel, sit, stay, and come level, is always imperative for general adjustment. Our experience with electronic bark collars and citronella collars convinces us that either can be used humanely, with dogs usually making the connection between barking and punishment quickly. Either collar is a far more reasonable solution than eviction. Efforts to screen city noise can also be made, and it may be helpful to leave the radio or TV turned on. Boredom and loneliness can be alleviated by a program of roadwork, exercise, massage, grooming, and proper diet, as outlined later in this book. The possibility of providing a companion for the dog might also be explored. Don't automatically think in terms of another dog — a cat may be a possibility, if the two are compatible. A bird that sings or can be taught to talk is another possibility.

Dogs who must face long periods of time alone should be greeted and left calmly. The owner should not make good-byes dramatic or prolonged, pleading with the dog not to chew or bark. When the owner returns, the dog should be greeted simply but affectionately. Overdramatic hellos and good-byes often keep dogs on edge and can result in stress-relieving behavior such as destructive chewing. After the owner leaves, the dog is still excited from being petted and cuddled, and possibly pleaded with to "be good." The owner may leave feeling better, but the dog may be on the verge of emotional collapse. Greeting and leaving scenes must not be the high points of the dog-owner relationship.

To burst into the house or apartment laden with special treats and then effusively greet the dog may alleviate some of your own guilt over leaving him isolated, but it's a disservice to the dog. The dog's psychological alarm clock tells him when to expect you home. The dog gears himself up for the happy moment, the treats, the play session. If, by chance, you are late, as is often the case because of subway, bus, or traffic delays, the dog's anticipation can turn into frustration, and frustration into destructiveness, whining, or barking.

If you live in a city and experience any of these problems, immediately begin to reconstruct your hello and good-bye scenes. Obedience training will help you to gain a leadership role over the dog. Even-keeled hellos and good-byes should give the dog a sense of purpose. For instance, say, "Watch the house," or some such phrase as you leave. When you return, praise the dog with a cheery hello but don't fall all over him. When you leave, offer the dog his favorite toy. When you return, delay feeding the dog for half an hour or longer. If you return from work at 5:30 and feed the dog right away, you are helping condition him to expect food at that time. Then he is frustrated whenever you are delayed and arrive home later.

## Selecting a Dog for City Life

If you are about to select a puppy or older dog for life in the city, you should seriously consider the personality traits of different breeds. Breed traits, though not absolute indicators, give you reasonable guidelines from which you may then narrow your focus. Size is not an automatic disqualifier. For example, it might appear that a German shepherd dog is ill prepared for city life whereas a poodle would do well. But this is not always the case. A happy life in the city depends on the individual dog. Conscious of the rising need for dogs as protectors and companions in urban areas, many breeders are selectively breeding dogs that can take city stress and adapt to the urban environment. For instance, German shepherds from certain bloodlines can adapt well to city life, but others cannot. Some poodles may do well in a city environment, but others do not. It's a good idea to talk to a breeder who is breeding for pets with a high threshold for noise, low excitability, and high trainability.

On the other hand, breeders can manipulate genetics only to a certain extent. Borzois are hounds that love to gallop, so they will always need an opportunity to run, which may be hard to find in the city. Malamutes, Siberian huskies, and other northern breeds may never adapt to the summer heat in a busy metropolis or be able to resist digging an occasional cooling hole. While individuals within a breed may adapt well, breed characteristics should still play a role in your selection of a city pet. As we have indicated, don't be fooled by size. Though the Doberman is a hefty dog, most Doberman bloodlines produce excellent city dogs. The Pembroke Welsh corgi, technically classed as a working dog, is "apartment size" but very active and may need an extraordinary amount of exercise.

To select a city dog properly, first decide objectively on your breed. Next, try to find someone who has a dog of that breed in your city. This may take some searching, or it may be as simple as stopping to chat with someone in a park or someone who happens to be walking the kind of dog you want. Most national breed organizations are happy to refer you to breeders specializing in city pets or to owners of urban dogs. It is well worth your time and trouble to meet the owner of a well-adjusted city dog and talk over breed characteristics and potential troubles. Most breeders and dog owners enjoy the opportunity to talk about their dogs.

It is possible to have a happy, healthy dog in the city — but it takes twice as much dedication. There is a tendency to look at dogs in terms of the services they render their owners. Whether that service be protection, companionship, or an aid to status, it is always secondary to the quality of the dog-owner relationship. The dog must feel responsible to, not for, his owner. He should perceive the owner as a helper and leader. The owner should act as the alpha figure in the dog's life. If the dog-owner relationship is marked by affection, regard, and love, the dog will reciprocate with characteristic steadfastness, respect, and friendship — regardless of where he lives.

# 16

# Suburban Life

The suburbs may be the best of dog worlds, but its environment poses its own special set of problems. Although suburban dogs are usually not as restricted, regulated, isolated, and controlled as those in the city, the very lifting of these restrictions provides a set of pressures for the suburban dog owner.

When suburbs do have leash laws, residents do not always obey them. In general, enforcement of leash laws is lax in the suburbs. The law may include a stipulation that the dog must be leashed or "under the owner's direct control." Having a dog under one's "direct control" is, of course, a vague concept. What it may mean in practice is that the dog is allowed to run free but eventually returns home. This is enough "control" for some owners. Free-roving dogs often form packs or bite, a growing problem in many suburbs and villages. This is forcing suburban municipalities to adopt city-type leash laws and implement zoning restrictions that penalize all dog owners. At least in the city, most stray dogs are picked up promptly and impounded. As a result, city dog owners tend to keep their dogs supervised, since they stand a very real chance of losing them otherwise.

Regardless of the environment, the only complete solution to free roving is to somehow contain the dog on one's own property when she is not on leash. The best and most humane way is to bring the dog into the house, where she belongs — regardless of the owner's interpretations to the contrary. Assuming the dog is indoors at least 50 percent of the time and is obedience-trained to come when called, there will be little or no problem of her going off the property. If there is a problem — or even as a preventative measure — additional backup solutions would be fencing in the entire yard, setting up a smaller fenced-in area, or rigging up a cable runner between the house and a tree in the yard. The latter is preferable to chaining the dog, since the runner allows the dog much more mobility. Nevertheless, it is imperative to use these as support to an overall relationship with a dog that includes regular exercise with the owner. One of the most frequent misjudgments suburbanites make is to think that their dogs get enough exercise in a fenced-in backyard. "But she has the whole backyard to herself during the day," we've heard many an owner complain when we recommend additional exercise. Dogs often sleep or are inactive when owners are away, making periods of walking and exercise important ingredients to the health of your relationship.

Some suburbanites persist in believing the myth that their environment is "countryish" enough to allow their pets to go where they please. (Unfortunately, even a country environment does not allow that.) The suburbs are not the country, and even if they were, that is no excuse for letting a dog run wild.

Many a suburban dog owner experiencing house soiling, chewing, digging, or free roving has asked us if we would like to adopt a mascot for our monastic community. Aside from the fact that we already have enough dogs of our own and are responsible for ten, sometimes twenty, boarding dogs, a country life here is not the solution to the dog's problems. Dog owners cannot do without having complete control over their dogs. A dog will come when he is kept close by, oriented to the inside of the house, and formally practiced in coming when called. Keep your dog inside, and either accompany him under supervision or leash him for defecation and exercise. If you want him to run free (presuming you have trained him sufficiently in the recall), take him to a park or large field and personally watch him. If your dog is not yet trustworthy off leash, a long fifty-foot clothesline can trail behind as the dog chases after a ball or Frisbee, thereby allowing you to deal with unexpected distractions or game playing.

Another common suburban dilemma occurs when a dog is left alone all day while the owner works. These dogs are left either inside or out, and quickly develop such problems as overbarking, frustration chewing, or fence jumping. Often they are enticed by other suburban dogs who are not restricted. It is difficult for a male dog to resist fence jumping if a constant parade of females in season passes his way while naive owners think their females are out doing their business. Some suburban dogs stay in yards that are fenced-in but afford a full view of a neighboring dog's yard. The result can be a virtual daylong barkfest between two or more of these animals. Remember, if one dog can see another but can't get to him, barking or whining at the other animal is the usual result. In fact, regardless of whether any other dog is in sight, any arrangement for keeping the dog that includes barrier frustration (cages, pens, chains, clothesline tethers) runs the risk of producing overbarking.

The installation of a dog door, giving the dog access to both the inside and outside, often resolves these problems. If a door cannot be installed, try to screen the dog from disturbing stimuli, whether they be other dogs, traffic, or passersby. Ideally, a good dog should be able to stay in the house without exploding at every different sight or sound. If not, it's the owner's responsibility to seek out training to help the dog cope.

Clients often look amazed when we suggest a dog door. Won't that encourage burglars? they ask. Possibly, but a window or regular door can encourage thieves, too. A small opening obviously meant for a dog does not entice most burglars. Naturally, they are wary of households with dogs that can bark, bite, or otherwise call attention to their presence, and the dog door advertises that you have one. With a dog door, the dog can defend both inside and outside areas. The expense of a dog door is minimal compared to its benefits. Some models can be installed even on rented properties, since the cutout section of door or wall can be reinserted. This can even be done in a sliding patio door opening.

There is nothing wrong with chaining a dog for a short period of time for elimination, but leaving a dog on a chain all day is bound to produce undesirable results. A chain is a last resort. The ideal suburban setup, which can be used if the owner is home or away, is a small enclosed pen, preferably connected to the interior of the house by a dog door. The floor of the pen should be grass or gravel. Converting a concrete patio into a dog yard is all right, but concrete can give the dog trouble with his pads, so check them often if your dog stays on concrete for any length of time. Be creative in constructing your pen (see chapter 27 on canine incarceration).

Some suburban dog owners experience trouble with many of the dogs in their immediate area. A highly successful approach is to start a local obedience class together using this book and other references. (For those interested, Winifred Strickland's *Obedience Class Instruction for Dogs* is specifically geared to this issue.) In this way, area dogs come to recognize and respect all local adults as alpha figures, and the trouble that ganged-up or individual suburban dogs can get themselves into decreases. If you cannot secure the cooperation of other dog owners in such a project, you can at least talk to them about the value of keeping their dogs supervised. The old saying "Love me, love my dog" applies here — you will have to be tactful in discussing dog problems with neighbors.

In suburbs where there are not enough fences and far too many dogs, fighting becomes a frequent event. Since many dogs are jammed together in a small area, each dog's territorial boundaries are frustrated. The only sensible approach is for all dog owners to obedience-train their dogs, restrict defecation to the area surrounding the house, and limit free roving.

A suburban setting can provide a wonderful life for a dog. But the owner must keep in mind that many other dogs rub shoulders with his dog. The suburbs are not the country, and at any rate, the country is not the answer to all canine problems, as we shall see in the next chapter.

# 17

# Country Life

There is no doubt about it: dogs do like life in the country. Each year thousands of dogs go on vacation with their owners. They travel to the seashore, to the woods, to the mountains. Many clients who live the greater part of the year in the city have told us about the amazing transformation that comes over their pets once they settle into the summer retreat. The fresh air, open space, and freedom from normal restrictions can do wonders for a dog. Dogs who live in the country all year enjoy an environment that closely resembles the milieu of their wolf ancestors.

But canine country life can have its problems, too. Many have to do with the fact that although life in the country is freer than life in the city or suburbs, the dog still must be responsibly restricted. The "myth of the country dog" states that a dog in the country is free to come and go as she pleases, need not spend any length of time indoors, and actually prefers to live outdoors. This may be true of certain rare dogs but is usually the owner's conception of canine liberty, not the dog's. Life in the country unfortunately allows owners to project their image of canine happiness on their dogs more easily. Often it is simply a rationalization of their own irresponsibility. Given that our monastery is located in a rural setting, we have seen many instances in which local dogs are simply allowed to run free. Often we hear reports of how this or that one was hit by a car or suffered some other sort of serious accident. We believe that in some respects country dogs are at greater risk because many owners seem to assume that they should be able to take care of themselves. Such an attitude puts your dog in danger.

Regardless of where you live, it is essential for you to spend time with your dog: not only to foster responsible management of your dog but to provide your dog with what he really wants — companionship. That should be priority number one for all dogs. We have heard many troubled urban or suburban problem-dog owners say, "What Fido needs is a family in the country that would take him — then he could just run all day." Trust us, things are not so simple. Even in the country Fido can't "run free all day" without getting into some kind of trouble and without having his need for companionship met.

William, a five-year-old Labrador retriever, and Duffy, his one-year-old son, belonged to a family that lived in the Vermont hills. The owner called us after both dogs had participated in killing a pig at a local farm, about three miles from their home. William was an old hand at this sort of thing and had been responsible for several other witnessed kills, including those of several geese. Duffy was relatively new to his father's occupation but had a history of free roving and chasing cars. Both dogs were left out in the morning and called home about five P.M. for dinner. Attempts to recall them began to falter, however, as the two dogs managed to find "dinner" elsewhere. They began to arrive home later and later, until they eventually stayed out all night. Like the proverbial drunk, they would arrive home staggering and exhausted. Both dogs would immediately go to the tree where they were usually chained for punishment. They would be chained for the day and night, at which point the owner would release them, and the whole ritual would be repeated. Asked why the dogs were allowed to run free, the owner explained, "Well, we live in the country, and I didn't think there was much trouble for them to get into."

Once William and Duffy were granted entrance to the house, their behavior changed remarkably. William refrained from leading Duffy into troublesome activities, and both dogs spent much of the day napping. Exercise, however, had to be regulated on leash, since the dogs had memorized a definite route during their adventures away from home and occasionally felt the urge to visit their old stomping grounds. After a few weeks of obedience training, the dogs could be allowed out into the yard and came back into the house when called.

Sarah, an eleven-month-old female Norwegian elkhound, lived in a rural area. Part of her owner's land abutted a sanitary landfill, a garbage dump. Sarah discovered the landfill when she was four months old, which is the age many puppies become more independent and begin seeking adventure away from home, if permitted to do so. Sarah loved the landfill and was observed by many motorists chomping away at leftovers and licking out empty tin cans. Even though her owners knew where she was, it didn't seem to bother them since, as they put it, "We want her to feel free!" Sarah's "freedom" soon led her into a bout with sarcoptic mange and a serious problem with fleas and ticks. Even after veterinary treatment, she was allowed to frequent the dump. About her eighth month, she began menacing other dogs that visited "her" dump and began growling at motorists who left their cars to empty trash. Threatening calls from the manager of the landfill convinced the owners to control the dog, but whenever she was left unsupervised, Sarah made for the dump.

Both of the above situations involve an owner's misunderstanding of the role of the dog in the country. The old misconception that country dogs are somehow entitled to run free is played out again and again, with disastrous consequences. William and Duffy were easily rehabilitated by a program of obedience training and in-house living. Sarah's owners needed to learn discipline techniques to correct her aggression.

In both cases, it took time for the scent of daily markings from the dogs' urination to die down. Dogs define their territory by spreading pheromones, chemical substances secreted by many animals, not just dogs, and used as calling cards when communicating with members of the same species. They are passed in urine, in feces, and possibly by breath. The frequent leg lifting of males is an attempt to mark out territory, which is later defended from invasion by other dogs and possibly humans. Though more common to males, pheromone-connected aggression can take place in females also. William and Duffy had successfully staked out land within a three-mile radius as their own. They considered the livestock in that area "theirs." Sarah no doubt perceived the landfill as an extension of her own backyard, and as she matured, she began to defend the dump as her territory.

Dogs who live in the country must limit their urination and defecation to the area immediately around the home, as they usually do in the suburbs. In the city, stools are regularly swept away, destroying the defining characteristics of pheromones. This probably eliminates a considerable amount of aggression among city dogs, who, nevertheless, are often on leash and under the control of their owners. But in the country a dog can mark off a large area of land and feel compelled to visit his territory each day to defend it from real or imaginary invaders. Add the possibility that an individual dog may have low discriminatory powers and high excitability, and you have all the elements for harmful aggression.

Unless the owner is able to structure the dog's life so that he leaves his canine calling cards only in his own yard, there is little chance of curing running away, aggression, or predation. In recent years one highly successful device for keeping dogs confined to property is the "invisible fence." This is an antenna wire that is buried in the ground around the perimeter of your property (or the area you desire to block off). A hidden transmitter carries a radio signal through the wire. Dogs wear a computer collar that picks up the signal coming from the wire and sets off a sound that indicates to your dog when he gets too close to the boundaries. Most companies that install invisible fences provide a training program that teaches your dog to understand the boundaries and stay inside the invisible fence. For most dogs the program is remarkably successful and humane. We find this device particularly helpful in rural areas, but less so in suburban settings that have a dense population of dogs, since it does not keep outside dogs off the property. Without question, it is the most successful and humane way of dealing with roaming behavior.

However, invisible fences can be pricey. If such a device is not a possibility for you, in addition to your restricting the dog's urine and feces to your yard, it is also necessary to discover exactly where your dog is traveling. Dogs who frequently run away usually have some place to go. Find out where. Inevitably the attraction is food, an opportunity to fight or play (or both) with other dogs, or the opportunity to breed. Occasionally some misled human will congregate dogs by passing out food. A simple phone call can stop the handing out of tidbits, but other attractions are harder to remove. Since it is usually not possible to remove them completely, a program of strict in-house living must be inaugurated, with defecation and urination on leash until the dog reorients himself to the home environment. Even so, that probably won't be enough. For starters, dogs that roam are often not in a subordinate relationship with their owners. To really begin to deal with the problem, this more general issue needs to be addressed. Training enhances bonding, which can help keep the dog on the property. Lastly, consider neutering your dog. Studies have shown that the most pronounced effect of neutering a male dog is to limit roaming, which is often a response to scenting a female in season.

If your dog is involved in predatory behavior, feasting on ducks, chickens, pigs, or even "big game" like deer or bear, your approach must be the same, with some further exploration. Predatory behavior is not always easy to prove, especially if the dog is working off his own territory. The dog may be involved in a pack, with two or three animals doing the killing and everyone taking part in the feast. Just because your dog has been sighted at the scene of a predatory incident does not mean he actually killed any other animal, but you should immediately contain the dog in his home area anyway. In most states a farmer has the right to kill any dog found molesting farm animals.

Predation that takes place in the home barnyard is more complicated. Owners who want loose dogs and free-roaming chickens to coexist are asking a lot. Chickens provoke chase and capture by dogs, with their flapping wings and cackling. Often a dog who would show no interest in a silent animal will enjoy a good chicken chase if the bird runs away and puts on a good show.

We have learned several lessons in our experience with predation cases. First, the old saying that it is harder to cure a dog of predation once the dog has "tasted blood" seems to be true. Rehabilitating these dogs often involves extensive work on a long line, in double-blind situations, in which the dog does not know he is being observed. This calls for the skillful timing and quick response of a capable trainer. If you have a chronic predator on your hands, a dog you can't seem to convince to eat dog food and nothing else, see a competent trainer as soon as possible. If, however, your dog has developed this habit recently, controlled setups and discipline that allows you to correct in the process of the chase are highly effective. Obedience training to the come, sit, and stay level is imperative to help strengthen the owner's leadership position, which is often weak.

Some trainers have used emetics effectively, lacing the "kill" with ipecac or other substances that cause vomiting, but we have not seen much success with this method. The use of electric collars to deal with predatory behavior, though promising, should be approached with caution. In recent years, electronic technology in dog training has become increasingly sophisticated, but it should not be looked upon as an automatic be-all, end-all solution to behavioral problems. It requires specific knowledge and skill to use such devices optimally. Seek the help of a trained professional who is familiar with such training and who can assist with your specific problem.

That said, most predators perceive of themselves as the leader of the pack, a basic misperception that must be cleared up quickly. The dog-owner relationship needs to be effectively reordered, with the human as leader, and the dog's freedom restricted to his immediate territory. In cases of home predation in which you catch your dog in the act, physical discipline under the chin or with the shakedown method can have a strong impact on your dog when followed up with set-up situations in which the dog is challenged to overcome the temptation. For example, a dog who loves to chase after cats or squirrels can be set up with the help of a fifty-foot rope. Have your dog in your yard with you at a time you anticipate such distractions. Be sure to wear gloves and keep your eye on your dog. Let him trail the rope around the yard. The second he spots the animal and begins to chase after it, grab the rope and run in the opposite direction. The correction will be meaningful and will begin the process of teaching him to ignore the distraction. Still, the best policy is to always think preventatively. Monitor the whereabouts of your dog at all times. Running away and predatory behavior are avoidable if the country dog owner simply keeps the dog nearby, or in a fenced-in area on the property.

An often successful and highly practical solution to the problems of country dogs is to put the dog to work. A great many breeds can be taught to herd sheep or cattle. There are books that describe how to home-train your dog for farmwork. Hunting dogs and pointers can learn hunting skills. If you don't have a farm, there are still small tasks your dog can do. Sled dogs can be harnessed and taught to pull children around for rides. Draft dogs like rottweilers can be harnessed to wagons and transport loads. Bringing in the paper, baby-sitting, and protecting the home are canine chores most dogs are eager to perform.

The city dog must learn to heel with precision and to mingle with strangers peacefully; the suburban dog needs to know obedience commands and take his place as a member of the family; the country dog, too, has his special role. The fact that he lives in the country does not mean he is on a perpetual vacation, free from the restrictions and duties other dogs face. Like any dog, he needs affection, training, a sense that he belongs and is wanted. In short, the dog needs to be and feel included, not excluded, whether he lives on Park Avenue or in Podunk.

# 18

# Outdoor Life

Much of the advice in this book suggests that the dog with behavioral problems be moved immediately into the house. We further suggest that owners wishing to enhance their relationship with their pets, even in the absence of behavioral issues, include the dogs as much as possible in their regular social lives. The sections on canine sleeping habits, digging, chewing, aggression, and other problems make it clear that we feel dogs belong in the owner's "den" (house or apartment) as an integral part of a "pack" (family group) under the supervision of a pack leader or "alpha figure" (the dog's owner[s]). Our own lifestyle with our dogs illustrates this belief.

What about the dog who lives outside? The first question we ask is, why? When we ask clients who are experiencing behavioral problems this question, their answers can often be dogmatic and curt:

"Because I want him outside, that's why."

"He likes it outside."

"That's where a dog belongs."

"He house-soils inside."

"He chews [or digs, jumps up, doesn't obey] if he's inside."

"I don't want hair all over the house."

"He bothers company, and we have a lot of company."

# 18

# Outdoor Life

Much of the advice in this book suggests that the dog with behavioral problems be moved immediately into the house. We further suggest that owners wishing to enhance their relationship with their pets, even in the absence of behavioral issues, include the dogs as much as possible in their regular social lives. The sections on canine sleeping habits, digging, chewing, aggression, and other problems make it clear that we feel dogs belong in the owner's "den" (house or apartment) as an integral part of a "pack" (family group) under the supervision of a pack leader or "alpha figure" (the dog's owner[s]). Our own lifestyle with our dogs illustrates this belief.

What about the dog who lives outside? The first question we ask is, why? When we ask clients who are experiencing behavioral problems this question, their answers can often be dogmatic and curt:

"Because I want him outside, that's why."

"He likes it outside."

"That's where a dog belongs."

"He house-soils inside."

"He chews [or digs, jumps up, doesn't obey] if he's inside."

"I don't want hair all over the house."

"He bothers company, and we have a lot of company."

"He needs the fresh air and exercise."

"We did it with our old dog, and it worked."

Occasionally there may be compelling reasons for keeping a dog outside on an exclusive basis. There may be an allergy in the family, or an elderly or handicapped person in residence. As any dog owner knows, certain sacrifices must be made to keep a dog. Hair on the furniture and floors is a reality of life with a dog, as is an occasional accident. It is a rare dog who needs exercise or fresh air twenty-four hours a day.

If your dog must live outside, provide her with a large yard and a good doghouse. The doghouse should be made of wood and painted a light color in summer and a dark color in winter. The color of your doghouse affects its interior temperature. The floor of the doghouse should be carpeted or otherwise insulated. Cedar shavings are good for warmth. Outdoor dogs need more fat in their diet when the temperature drops down into the forties or lower. The most convenient method is to add a tablespoon of vegetable oil to the dog's daily diet.

But most often a dog lives outside for one of two reasons: her owners prefer it that way and see no reason to change the dog's lifestyle, or the dog has been tried indoors and was too unruly. If the second reason is the excuse, we hope this book will guide you in coping with behavioral problems so that your dog can reenter your living quarters. If you simply prefer (or demand, as the case may be) that the dog live outdoors, there is probably no changing your mind.

Our advice to such owners faced with behavioral problems is to try the dog in the house for one week. Expect the first couple of days to be hectic. Let the dog sleep in your bedroom. If this is strictly out of the question, we suggest a limited program of obedience training done outdoors. Remember, no guarantees can be made about changing behavior that is related to social isolation, unless the dog is no longer banished. Owners who are absolutely intransigent about letting the dog into the house, even on a limited basis, might do well to consider placing the dog elsewhere and investing in a domestic animal that adapts more easily to life outdoors — for instance, a horse, a cow, or a pig.

# Sensitivity Exercises

# 19

# Your Dog May Be Lonely

We hear a lot about America's pampered pet population these days, but such pets are in the minority. Dr. Benjamin Hart, currently director of the Center for Animal Behavior at the University of California at Davis, remarks, "Ninety-nine percent of pets aren't pampered." And he adds, "If your dog is tearing up the place, he could be lonely or he could need more exercise."* Can dogs experience loneliness? Quite easily, suggests Dr. Michael Fox, whose research indicates that the "emotional centers" of the dog's brain are similar to human emotional centers.** While studying human loneliness, psychologist James L. Lynch conducted a series of animal experiments showing, among other things, that petting produced profound effects on the cardiovascular systems of both dogs and humans.† Even more suggestively, Rupert Sheldrake's recent study on the telepathic powers of animals makes a strong case that many dogs possess telepathic connections with their owners.‡ In a remarkable presentation grounded in serious science, Sheldrake shows that many dogs actually seem to know when their owners are returning home after lengthy absences. He comes to this conclusion from having conducted a number of

*Dr. Benjamin Hart, quoted in Associated Press dispatch, *Boston Globe*, August 1977.

**Dr. Michael Fox, *Integrative Development of Brain and Behavior in the Dog* (Chicago: University of Chicago Press, 1971).

†James L. Lynch, *The Broken Heart: The Medical Consequences of Loneliness* (New York: Basic Books, 1977).

‡Rupert Sheldrake, *Dogs That Know When Their Owners Are Coming Home* (New York: Three Rivers Press, 1999).

controlled experiments that used simultaneous videography (owner-dog) and a random schedule for the owner to return home. After considering the possible reasons for the data, his best explanation is telepathy. If true, it would clearly demonstrate that animals are emotionally connected to their owners even when the owners are not present and that they do experience loneliness.

This research merely confirms what professionals working with dogs have known all along. One of the biggest obstacles to healthy pet-owner relationships is pet loneliness. Dog owners, busy with their own activities, may never suspect that their friend suffers from isolation. A case in point: Sassy, an Airedale terrier, spent the hours between eight and five at home, alone. Her owners worked and there were no children in the family. She was purchased when she was four months old. Her owners were concerned that a younger puppy would not be able to adjust to long waiting periods alone. "Now we realize we should have purchased an even older, trained dog, used to entertaining herself alone," one of the distraught owners confessed. The ensuing conversation focused on how the age of the dog was to blame for the destruction the pet wreaked when the owners were gone. But after a week's observation, we noticed that Sassy responded well to four- or five-hour periods of isolation, entertaining herself with toys, napping, and looking out windows. She was not tense or anxious but became so after six or seven hours. We were able to observe the dog through a one-way mirror. Though her owners had complained of Sassy's lack of pizzazz and spirit, on our turf she was exuberant and playful.

In our next interview with Sassy's owners, we explored new areas. Sassy rarely left the house. She had been in a car twice, the first time when she was brought home from the breeding kennel, and the second time for a visit to a vet. The owners rarely had guests in their own home but were active socially and went out often at night, leaving Sassy alone for a second lengthy period of time. Questions about play periods, obedience training, fetch games, and roughhousing elicited puzzled stares from the owners. The situation was becoming clear: Sassy was lonely and vented her frustration toward the end of long periods alone. Because she had been conditioned not to expect play periods or extra attention, her sullenness and lack of animation had become a generalized condition.

We suggested a daily play session, with both owners on all fours. Advising against overly emotional hello or good-bye scenes, we

*It is vital to play with your dog. Here a rottweiler plays a retrieval game with her owner.*

nonetheless suggested a meal about half an hour after arriving home, followed by a walk through the neighborhood on leash. To bring Sassy into contact with more people, we recommended obedience training so that Sassy could be included in shopping trips, outings, and, if possible, parties and get-togethers with neighbors.

Sassy recovered quickly, and the destruction stopped. Weeks later the owners reported that they had arranged for alternating trips home at noon to take the dog for a walk. They had ceased socially over-extending themselves in an effort to bolster their business careers, and they now reserved two nights a week to stay home, including their dog in the family circle. They began to host parties, taught Sassy standard obedience work and some parlor tricks, and showed her off to friends. They stopped excluding the dog from their lives, firmly integrating her into their daily schedules.

Dogs are social animals, and they need to be included in a pack. Since we have deprived them of their normal pack — animals of their own species — and the freedom to set up social structures of their own, we must include them in our pack and help them adapt to human social structures. Because so many of us are simply out of touch with our own animality, and even more out of touch with the kingdom of animals, our initial reaction is to deny animals entrance to our human world. The old dichotomies of good and evil, body and spirit, animal and man, are still played out dramatically in pet-owner relationships.

Many pet owners perceive their charges to be incapable of enjoying human company. The tendency is to isolate dogs rather than include them. Although dogs are allowed in supermarkets and restaurants in many parts of Europe, they are barred from such places in the United States, where they are considered a health hazard. Shopping centers and malls now frequently forbid pets, on or off leash, and in some large urban areas, dogs are even being prohibited from city parks. The social situation for our pets promises to get worse, not better. The capital of Iceland and Roosevelt Island in New York, to mention two areas, now forbid ownership of dogs. Dogs will continue to be ostracized and isolated as long as the pet population soars and owners act irresponsibly. What all this adds up to, from the dog's point of view, is more isolation, more boredom, and more loneliness. This, in a creature that is genetically a pack animal!

But there are ways of providing an enriching communal life for your dog. First of all, don't leave your dog alone if you can help it. The built-in tendency is to leave the dog home, but stop and reflect: can I take the dog with me? You might be able to. We have a contractor friend who takes his dog Jennie with him to work every day. He has the dog trained to behave well in his car, and he takes appropriate precautions about parking in the shade and making sure she has enough ventilation and water. He is able to interact with her during coffee breaks and take her out for an exercise session at lunch, as well as interact with fellow workers. The benefit such contact has brought to their relationship is dramatic. Second, provide as much varied human interaction for your dog as possible. Obedience-train the dog to the come, sit, stay, and down level. Begin to take your dog to busy streets, shopping centers (if allowed), and other congested areas where your dog can observe people in motion, look at neutral passersby, and in general get a feel for what it is like to be around large groups of humans. If possible, extend this exposure to include family reunions, outdoor parties, and other situations.

## Keeping Your Dog Close

At New Skete we try to keep our dogs with us as often as possible. A puppy begins by simply following the Brother in charge wherever he goes. Most puppies have a natural tendency to follow humans, and we try to maximize this inclination. New puppy owners should include a ten- or fifteen-minute session daily when they have their pup follow them, off lead, while changing pace, swooping into turns, and keeping the pup animated and happy with high-intensity encouragement. This simple procedure helps alleviate come-on-command problems later in life. Puppies who are hesitant about following their masters should be leashed, with the leash attached to a belt

*Most dogs can be happily integrated into your daily schedule. They love to be close by and watch.*

loop. The monks here often proceed with work that requires both hands simply by tying the pup and leash to their belt loops.

Older dogs can be taught to follow and stay close using this same method. The older the animal, the more conditioning on leash is required before the dog gets the idea that staying close is a pleasant experience. Owners who complain that they cannot keep their dogs with them or take them along on outings because they will run away at the first opportunity should try a few days using the leash as an umbilical cord. Needless to say, other aspects of the dog-owner relationship must be in order if the dog is to learn to prefer the owner's company to any other activity.

To give you an idea of how successfully a dog can be integrated into a busy schedule, let's look at two daytime rituals. The first is the schedule of a monk at New Skete, and the second is a timetable for a busy woman we'll call Mrs. Bede:

## Monk's Schedule

5:00–6:30 A.M. New Skete monk rises. While he attends to personal care, dog is let into outdoor defecation and exercise run. While monk meditates and prepares for church, dog stays quietly in bedroom or in a kennel crate.

7:15 A.M. Matins (morning prayer). Brother goes to church; dog waits in monk's room or a kennel pen. Dog is fed and left alone to eat. This helps pass the time until the monk returns, when dog is let out again into the run.

8:00 A.M. Brother eats breakfast in monastery dining room. Dog lies down near dining-room wall until breakfast is finished. Afterward, monk may take a walk with dog.

9:00 A.M. Work. If possible, monk takes dog with him wherever he is working that day. If not possible, dog is kept in a large outdoor exercise pen with a playmate. For Brothers cooking or doing house chores, dog is put nearby on a long down-stay, which is enforced. Monk does not allow random running around. Work in the church can be done with the dogs lying down in the sacristy.

12 noon Monk eats lunch. While the Brothers eat, several dogs lie down nearby as at breakfast. No begging or coming to the table is allowed.

*Well-behaved dogs can be present at mealtime or at a snack break without becoming a nuisance.*

1:00 P.M. Afternoon work period. Same as morning work period for both monks and dogs.

3:00 P.M. Some Brothers continue working, while others take their dogs for walks, give them obedience lessons, or give them a siesta time.

5:00 P.M. Vespers. Brother goes to church; dog waits in room. Some dogs are fed again at this time.

6:00 P.M. Same as above for meals.

7:00 P.M. Community recreation. Dogs hold down-stays in a relaxed manner and are natural part of the communal atmosphere.

approx. 8:00–9:00 P.M. Brother retires with dog to his room to read, prepare for sleep.

On weekends, there are more church services and more free time. On Sundays after liturgy, dogs are socialized during a coffee hour with those who attended church.

## Mrs. Bede's Schedule

7:00 A.M. Rises, takes dog from her bedroom and puts him outside in enclosed yard to take care of his needs, prepares breakfast for children. Helps prepare for school.

8:00 A.M. Dog accompanies children to the bus stop in front of the house. Mrs. Bede calls dog back home immediately after bus departs. Dog eats.

9:00 A.M. Dog watches indoors as Mrs. Bede attends to housework.

11:00 A.M. Shopping trip. Mrs. Bede loads dog and two preschoolers into car. Dog stays in the car while she shops.

12 noon Dog lies by table as Mrs. Bede prepares lunch for preschoolers and herself.

1:00 P.M. Children nap. Dog takes nap in children's room at same time. Mrs. Bede naps or does ironing, other housework.

3:00 P.M. Two children return from school; dog instinctively goes to door to be let out to meet bus.

3:15 P.M. Children and dog return from bus stop, have snack in kitchen. Children have play session in backyard with dog. After half an hour dog called in and confined.

4:30 P.M. Mrs. Bede gives dog fifteen-minute obedience session in backyard, includes children in training session.

4:45 P.M. Mrs. Bede prepares dinner with dog on down-stay in kitchen.

5:30 P.M. Mr. Bede returns home, goes jogging with dog.

6:30 P.M. Family eats, with dog on down-stay near table.

7:30 P.M. Dog recreates, watches TV, etc., with family.

10:45 P.M. Dog taken outside to eliminate.

11:00 P.M. Dog retires in master bedroom for the night.

Notice that both schedules keep the dog near his master. The dog is included, not excluded. The dog is often with people, although one or two private rest periods are included in both schedules, timed with human rest periods. Similarly, when humans are taking care of their personal needs, dogs are, too. The dogs are left alone to eat, in contrast with the "Grand Central Station" atmosphere of many pets' eating time. Periods of exercise are included. The dog is treated as a true companion and friend, part of the family circle. The dogs in these schedules are not treated as emotional cripples who need attention every minute of the day. This is particularly important to keep in mind with retired couples who may be caring for a dog. Because they normally have more free time than younger adults, the tendency is to go to the other extreme, whereby the dog is always with his owner, having no time to himself. If not handled reasonably, so much time together can result in dependency that makes kenneling and other necessary separations traumatic. Be sure to schedule some time each day when your dog is alone. On the other hand, these dogs' masters do not fall victim to the common dog myths we have already touched on — for instance, the myth that a dog needs to run free 80 percent of the day.

To prevent canine loneliness and the possible destruction and neurotic behavior that can stem from it, integrate the dog into your schedule. Take time to map out your day, seeing how you can include the dog in it. Don't assume that your dog is automatically eliminated from certain activities or areas — ask, inquire, and train. Some owners may find it possible to take their dogs to work, if they are obedience-trained and quiet. In the best dog-owner relationships, isolation is usually the exception, rather than the rule.

# 20

# Where Is Your Dog This Evening?

Where does your dog sleep at night? If your answer is "in the bedroom, on the floor," you probably already know the gist of this chapter. If your reply is "in the cellar," "tied in the kitchen," or "in bed, with me," read on. We will discuss the value of "sleep therapy" for you and your dog, and how to go about it.

One objection clients have when we suggest they have their dog sleep in the bedroom involves what they conceive of as the impropriety of the situation. One client put it this way: "I tie her in the kitchen at night. That's where she's always stayed. Sometimes she'll chew overnight. Until I got a steel tether, she used to chew right through the leather one until she got free. Then she would run into the bedroom. She would creep in, and I would discover her the next morning. So I began to shut the kitchen door. Then she learned how to open the latch on the kitchen door. So I shut the bedroom door. She began scratching on the bedroom door. What does this sound like to you?"

"It sounds as if she's trying to get into the bedroom," we responded. "Did you ever consider letting her sleep in the bedroom?"

"Heavens, no! My husband would never allow it. We might be in the middle of something! [Client clears her throat.] It just doesn't seem proper. But how am I going to teach her she belongs in the kitchen?"

Hopefully this chapter will help the couple above, and others like them, to get over a phobia about having a dog in the bedroom overnight. However, if you are absolutely determined that the dog stay out of your bedroom, perhaps you can provide comfortable alternative sleeping conditions. Though we highly recommend letting the dog sleep in the bedroom, we can see how it can be a bad experience if a nervous owner is sending out negative vibrations all night. The dog will pick up on them. But if you can see the value of the experience, from the dog's point of view, and are willing to try it, you will be surprised how fast your phobias fade.

## Bedroom Deportment

Your pooch is in the room and you're ready to retire. Although it is best to have trained your dog to lie down in advance (see chapter 35, "The Down"), it is surprising how even the most hyperactive dogs tend to plop down as soon as the lights are shut off. If your dog paces, runs around, or gets up too often, you might want to start teaching the down in advance and then tethering the dog to the foot of the bed. A good way to start this training is during the day, by taking (or appearing to take) a brief nap. That way, if the dog is restless, you can work on the down without losing valuable sleep. Begin with your dog next to the bed on a down. Lie down on the bed and relax. Often the

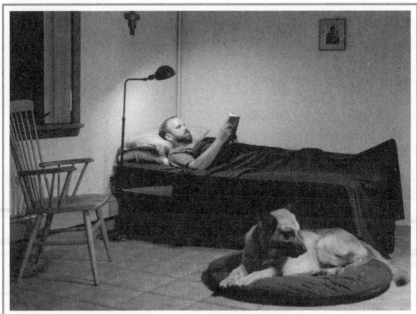

*Sleeping in your bedroom is an ideal way for your dog to bond with you.*

dog senses the shutting-down mode and relaxes as well, but if not, you are easily able to enforce the down. Be consistent. The tether keeps the dog from wandering around, even allowing you to correct the dog without getting up. By keeping the sessions short, you can progressively work on the procedure over a couple of days; before you know it, the dog will be perfectly at peace in your room. Allow no jumping on the bed or other horseplay. Discipline such behavior by curtly removing the dog from the bed and literally depositing her on the floor with a sharp "no!"

Provide a spot for the dog with a small rug, dog bed, or blanket. Food and water are not needed overnight. The best policy is to have the dog lie down, ignore her, and go about preparing for bed. Do not make a fuss over the dog. This is a time of quiet, uncomplicated interaction between you and your dog. It is a time when you let the dog into your private "den" but not to disrupt it. Most dogs simply find a wall and lie down against it. Some like to lie under a desk since it provides a denlike atmosphere. As long as it is not inconvenient for you, let the dog pick the spot. Don't force the dog to lie anywhere. If there is an area the dog lies on that bothers you, block it off by rearranging furniture or placing an object in the way.

Finally, turn off the lights. This is usually the final sign-off. If you have problems with pacing or hyperactivity (more common in males than in females), try turning off the lights and telling the dog to lie down. Most dogs circle in a holding pattern for a few revolutions, then land for the night.

## Get Up!

The canine privilege of inhabiting the master's den is just that — a privilege. Don't allow your dog to abuse it. If the dog bothers you during the night, give him a slight tap on the muzzle and the command to lie down. Pawing the bed or trying to get on it should result in slapped paws and a shove off. In general, it is a good idea to keep the dog away from the top of the bed but still in the bedroom. Dog owners who enjoy a long good-night scene or ritual bedroom romp are inviting trouble and canceling out the potential benefits that stem from an in-bedroom sleep. Don't go overboard. Allow your dog to share your den but not take it over.

If your dog bothers you during the early-morning hours, he may be obeying his psychological and physical alarm clock that is telling him to defecate, especially the younger he is. If your dog has a regular schedule for elimination, you need to stick to it, regardless of whether you want to sleep in. Many a dog owner has become an expert at stumbling to the door on a Saturday morning, letting the dog out and back in, then falling back into bed for a little more shut-eye. Part of being a dog owner is respecting your dog's inner schedule. The more mature the dog, the more control he has. With time, you should be able to sleep later, and the dog will, too.

On the other hand, some dogs take an invitation into the bedroom as a chance to play "doorman" with their owners. Don't get up "every hour, on the hour" for your dog. Assuming that your dog receives adequate exercise and is not restless simply because of excess energy, take the dog out before retiring and make sure he has had enough time to eliminate. During the night, ignore whining and shush him with a stern "no." In the case of young puppies, you should get up and take the pup out for a time (usually until he shifts to two meals per day), but older dogs can usually "hold it." Don't rev the dog up before bedtime, since this might encourage vomiting or defecation.

## The Value of In-Bedroom Sleeping

Of all the training exercises described in this book, sleep therapy is the easiest. You don't have to do much except let the dog in and out of the room and keep the whole experience in the room as low-key as possible. But this is a time when a lot is happening, from the dog's point of view. The dog is enjoying an extended period of time with your scent. The bedroom contains the most intense scents. Dogs focus on the bed itself, especially the center of the bed. Therefore, we prefer that the dog not be allowed on the bed. Aside from potential problems with dominance issues, owners who allow their dog on the bed may discover the middle of the mattress chewed up. Some contact with the owner's scent is beneficial; too much contact backfires. The rest of the bedroom is a fragrant delight. The closet houses shoes and socks, the rug on the floor is walked on by bare feet, and the drapes are touched constantly. For your dog, the in-bedroom sleep is a scent "high"— but a high that must be properly regulated and controlled.

Night is a time when the owner and dog can have extended contact without demanding anything from each other, a time when most dogs make their own decision to lie down and relax with you, shut down, turn off, sleep. This, in itself, while deceptively uneventful, builds trust and confidence between owner and dog. Consequently, the in-bedroom sleep can be a great help if you are experiencing problems with hyperactivity, social isolation, lack of rapport, night barking and whining, or general unruliness. For the owner without much time for a pet, it can be a final moment of contact and attention.

In the hundreds of "problem dog" cases we have worked with at New Skete, 80 percent of the pets slept outside of the bedroom, usually in the living room, basement, outdoors, or, significantly, just outside the bedroom door.

All of the New Skete monks keep their assigned dogs in their rooms at night. The bedrooms are in the cloistered section, where we have a rule forbidding any noise. Our dogs glide into the room behind us; then we give the down command with a hand signal and go about the business of retiring. Our dogs are usually asleep before we are.

A good example of the power of in-bedroom sleeping is what happens when we have an imported dog arrive from Germany. Since we try to include the very best bloodlines in our breeding program, some dogs must come from overseas. They may or may not know any English when they arrive, and are generally disoriented. But after a week of sleeping in the same room with one of the monks, they calm down and follow that Brother around like a shadow. This is a good tip for trainers and others who must get to know strange dogs in a short period of time in order to train them or nurse them — move them into your bedroom for sleeping, and wonders occur.

## Dog Dreams

Dogs do dream. They are often quite verbal about it, moaning and purring during the dream. Some owners have mistakenly thought it best to wake up the dog and stop the "nightmare." But the old adage "let sleeping dogs lie" applies here. If the dog's REM (rapid eye movement) sleep is disturbed too often, daytime hyperactivity and unruliness can result. If the dream becomes really noisy, try stopping it by crinkling a piece of paper or tapping the floor. This changes the pattern of the dream but does not wake up the dog.

To us, it is important where your dog sleeps. The best place is in your bedroom or in your den. Needless to say, the protection potential of any pet increases insofar as he can quickly alarm his owners of any danger. If the dog has access to your bedroom, you have a built-in burglar and fire alarm.

# 21

# Playing Pavlov

The Russian psychologist Ivan Pavlov (1849–1936) used dogs extensively in his experiments with conditioned reflex. Though he was not specifically concerned with dog training, he left a body of work that can be of great value to breeders, trainers, and dog owners. This chapter does not presume to explore all of Pavlov's work, much less in detail. Nevertheless, we do explain a canine socialization and training technique that is highly Pavlovian.

A simple set of keys can help you deepen your relationship with your dog and alleviate many forms of problem behavior. Dogs with recall problems, a tendency for chewing, digging, or other destructive antics, or appetite problems can be successfully "keyed in" by a simple form of sound conditioning. You need four or five keys on a

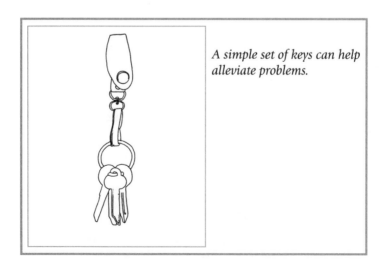

*A simple set of keys can help alleviate problems.*

key chain. Every second key should be brass, while the others should be made of another metal, preferably not aluminum. Brass and steel make higher-pitched sounds, and four or five keys sound better to dogs than ten or twenty.

The basic idea behind this sound conditioning is to precede desired behavior with a distinctive sound. Keys are used here because they provide a strong, high-pitched sound that is irresistible to the dog once she is properly conditioned. Hand clapping, whistling, and cooing are in another realm, obviously proceeding from a person. To these specifically human sounds, a dog may or may not respond, depending on the current state of the relationship between the dog and the person making the sound. Keys or whistles are neutral, and therefore more effective. In our experience, however, we find the most effective sound device to be the keys on a chain.

Let's take an example. Your dog doesn't come when called. If you have a puppy or a dog younger than two years old, your chances for effective sound conditioning to correct the "come problem" are better than with an older dog who is used to going the other way when called. Yet it is never too late to try this training technique. For the greatest success, you must have regular feeding times for your dog twice daily, and he must finish eating in about fifteen to twenty minutes. If you have your dog on the "nibbler plan," you need to switch to a regular feeding schedule and remove the food if it is not finished promptly. (We suggest this method of feeding in general.) Before placing the dish within his reach and allowing him to eat, get your dog's attention and jingle the keys for two or three seconds. Then go about your business as he eats. Do not make a show out of it, and preferably do not allow your dog to see you jingle the keys. You may attach the keys to your belt loop with a snap belt. Repeat this procedure at the second meal, continuing it for two or three weeks. Do not use the keys around your dog for any other purpose until you have spent some time doing this conditioning procedure whenever your dog eats. Another positive booster and "reinforcing effect" can be gained by using the keys whenever you return from work, from an errand, or in your car — stop the motor, open the door (both distinctive sounds in themselves), jingle the keys, and call out the dog's name in a happy voice that carries.

After two or three weeks, begin a daily session in which you call your dog, jingle the keys, and praise him lavishly while offering him a treat when the recall is good. Make sure you are crouching down, have a smile on your face, have your arms open to "funnel" the dog in to you, and are not overjingling your keys. The whole point, obviously, is to let the dog hear the sound of the keys — so that he reacts positively, on his own. Let the dog win. Then reverse the procedure by jingling first and then calling your dog's name. Finally, try it with the keys alone. Above all, make sure your praise is animated, verbal, and physical when your dog comes to you. There is hardly a moment in training more deserving of praise than the successful recall. (Again to reiterate: never call a dog to you and then discipline him. If you have anything other than a happy reception planned for your dog at the end of a come-in, go get the dog yourself.)

Conditioning with sound for recall is a training technique that builds with repetition. Expand the practice to other happy occasions, such as treats, play sessions, or car trips. Always follow up the key jingling with a happy experience and praise.

Key therapy works on a group level, too. Puppies are often sound-conditioned in lieu of giving them names, since the breeder may prefer to leave that prerogative to the future owners. As it is so effective with puppies and younger dogs, keys can help teach pups the recall, can aid in gait evaluation, and helps keep a litter grouped together and diverted from traffic and other dangers. One breeder detoured a litter of nine puppies from a busy road with a few jingles of her trusty keys.

We have long used keys with our puppies here at the monastery with great results. By starting the conditioning at feeding time, puppies become focused on the sound of the keys in no time, making preliminary training for the recall a snap. When a group of pups is out in the yard, investigating and playing with one another, a simple jingle of the keys brings the whole litter running toward the source, and to lavish praise.

We progressively extend this procedure to walking them around the monastery grounds and up and down steps, thereby also conditioning the pups to a variety of environments and surfaces.

Some professionals use the key method to double-handle dogs in the show ring. In double-handling, one person actually shows the dog, while the other provides ringside encouragement, often with strategic key jingling. However, we should note that this is a technique forbidden by the American Kennel Club, though it is the subject of some debate and practiced widely.

For dogs with appetite problems, begin as above, jingling your keys a couple of times before placing the food before your dog. Pavlov's dogs salivated at the sound of the bell, and the principle here is the same. Appetite problems often have many roots (see chapter 11 on feeding), so if your pet is already playing a food game with you, make sure that you are not simply adding the key ritual without first correcting the basic trouble.

Chewers, diggers, and general house wreckers can sometimes be diverted from their destructive activities by strategic sound therapy. Verbal or physical discipline is the usual corrective for this kind of behavior, but sound diversion may work for the dog who fails to respond to these methods. Keys, whistles, and other sound devices are even helpful with dogs who chew or dig when the owner is absent.

Here is just such a case.

Thunder, a two-year-old malamute, spent the day digging holes in the backyard. When the owners returned, it was standard procedure to wallop the dog and then isolate him. The digging, however, continued to get worse until, in exasperation, they went to their veterinarian, who referred them to us. Thunder appeared to be a normal, sound male, alert and active, with some holdover behavior from puppyhood. We combined a program of basic obedience training, some diet changes, and sound conditioning. Later, when Thunder left us, his owners began to condition the dog with a simple key set at mealtimes, early morning, and night. One of the owners had a day off and spent the day at a neighbor's house, which had a view of her backyard. Anticipating Thunder's digging, she jingled the keys several times that afternoon. Balls, sticks, and toys had previously offered no diversion to the digging. Construction of a simple obstacle course of old tires and boards and some large tree limbs did. When she returned to work the next day, she left the keys with her neighbor, who agreed to watch for signs of canine excavation. This helpful neighbor jingled the keys whenever she felt the dog was about to begin digging or if she actually caught the dog in the act. By the end of two weeks' time, with the combined impact of many facets of her program, the digging had stopped completely.

The owner next installed a dog door to give Thunder the option of being inside or out. Previously, she had been against this, fearing indoor destruction more than the outdoor variety. Later she reported that neighbors had observed the dog spending about 50 percent of his time indoors, with no destruction of any kind, in or out.

A simpler problem is exemplified in the case of a two-year-old German shepherd, Abbey, a lovely bitch of show quality. She was being campaigned to her championship, having already completed obedience work to CDX (Companion Dog Excellent) level. She had no trouble mixing the different types of work demanded in the obedience and conformation rings, but she had one slight problem. When anyone proffered a camera to take a show shot, her ears went down, her tail flipped inward, and her expression became dull and forlorn. Here, too, sound conditioning won the day. In all such picture-taking sessions, a second person would stand before her out of camera range and "bait" her with the keys; Abbey's whole demeanor would change, resulting in highly animated, attractive photographs.

However, keys (and in the following case, whistles) are not a magic solution to any and all behavioral problems. In the original edition of this book we told the story of Shana, a year-old collie mix who barked incessantly and was reconditioned with the help of keys to remain quiet in the absence of the owner. This case study, in which sound conditioning worked, admittedly involved a complex set of measures (including the help of several neighbors) that often is not practical in real life. We've also heard from some clients who have tried the same approach with their dogs without consistent success. As a result, we would approach this problem differently today. In addition to the general recommendations we made of initiating basic obedience training, toning down emotional departures and reunions, and making sure that the dog received plenty of exercise, we would recommend a bark collar instead of using keys and timed corrections. Bark collars today are humane and work very well: almost all dogs quickly make the connection between their barking and the correction. Another advantage is that the collars correct consistently. They do not depend on split-second timing and do not associate any physical correction with the owner (something that in this case seems preferable).

## Clicker Training

We would be remiss if at this point we did not mention a new type of dog training that has a strong relationship to sound and is becoming increasingly popular in some sectors of the dog-training world: clicker training. This type of training is based entirely on positive reinforcement and uses the principles of operant conditioning, what one of the chief exponents of this method, Karen Pryor, describes as "a set of scientific principles describing the development of behavior in which the animal 'operates' on the environment, instead of the other way around."* Translated into layman's terms, this means rewarding a particular behavior so that the animal will repeat it whenever the trainer desires. Before it was applied to training dogs, this approach was most famously used with dolphins, an animal you really cannot discipline. Furthermore, dolphin trainers discovered that corrections were unnecessary since the principles of positive reinforcement alone could be used successfully to train dolphins to work. By using toots on a whistle and a bucket of fish, trainers could shape dolphin behavior in highly refined ways.

Granted, there is a real difference between dogs and dolphins. Dogs live with us in much more intimate ways than do dolphins, making the relationship far more complex. Nevertheless, clicker trainers believe that the same positive principles can be applied to dog training in a way that makes the use of force obsolete. This is where all the controversy about this method lies: since everything is built upon positive reinforcement, there are no corrections or discipline whatsoever in the learning process. In fact, they are positively discouraged. Many wonder, "Is it realistic and appropriate to train dogs using this method alone?" Clicker trainers insist that it is and have spent enormous energy trying to sell it to the public.

Entire books have been written on clicker training. Without trying to overly simplify it, here is a brief overview of how it works: Clicker training is based on the presumption that a dog will always work willingly and efficiently, as long as he both understands and is motivated to perform the task requested. What is usually used as the primary

*Karen Pryor, *Clicker Training for Dogs* (Waltham, Mass.: Sunshine Books, 1999), p. 1.

motivator, or "reinforcer," is food, but theoretically it could be anything the dog really loves. Clicker training takes its name from a small, rectangular child's toy that is held in the trainer's hand and makes a sharp "clicking" noise when depressed. As a particular sound, the clicker holds no special fascination for a dog. However, it can become an "event marker," a "conditioned reinforcer" the dog comes to associate with a desirable reward (a treat, the primary reinforcer) received after performing a correct behavior. The click occurs at the precise moment the dog follows the command, and he then receives the reward several seconds (or many seconds) afterward.

The desired behavior can be progressively refined, or "shaped," because the dog wants to earn the reward. When a dog hears the click, he knows a reward is coming. Thus, the click becomes a positive reinforcer: the dog associates the click (and a verbal cue) with a specific behavior. By patiently teaching the dog the basic commands, a trainer can establish "fluencies." By now, you can see that clicker training has a vocabulary all its own, grounded in the language of operant conditioning. This can be a bit of a turnoff, but clicker trainers insist that it is useful to stick with terms that can be scientifically described and made precise. Hence, they define a *fluency* as "a basic, learned behavior that you can incorporate into other, more complex behaviors and into behavior chains."* These exercises are taught with the help of a target stick. (In clicker training a stick replaces the leash. Clicker training is done primarily off leash.) By building upon already learned exercises, clicker trainers believe the sky is the limit as to what you can teach your dog.

Clicker training has evoked strong feelings in the training profession, both pro and con. People seem to be either rabidly for or against it and are not averse to hurling stinging insults at the opposing position. Traditional trainers are billed as force fiends addicted to punishment, while clicker trainers are described as unrealistic sentimentalists who have been sucked into what is politically correct. So who's right?

Clicker training is not the method of training we specialize in, and although we have become familiar with it, we by no means pretend to

*Morgan Spector, *Clicker Training for Obedience* (Waltham, Mass.: Sunshine Books, 1999), p. 45.

be experts. Though we are not advocates of clicker training, at the same time we do not deny that someone can train his or her dog successfully using this method. The fact is that there are a number of highly skilled dog trainers taking this approach. Authorities such as Gary Wilkes and Morgan Spector, to name only two, have established reputations as highly competent trainers with a sound knowledge of dogs and dog behavior.

What we do take serious exception to, in all the rhetoric, is the suggestion that training *has* to be entirely positive for it to be humane, that corrections have no place in a modern thinking approach to obedience training. This is simply not true. Dogs can learn quickly and effectively with the help of well-timed corrections in a program of overall obedience that is essentially positive. Dogs are quite capable of understanding that there are negative consequences for certain behaviors beyond simply not receiving a treat, and this knowledge can lead to consistency and reliability in training results. There are simply too many well-adjusted, healthy dogs around that have been trained intelligently by such methods to suggest otherwise.

Furthermore, even though we are strong advocates of a positive approach to training, we do not believe it is either fair or helpful to suggest that corrections are never appropriate. In today's training climate, a lot of owners have come to feel guilty about giving any sort of correction to their dogs, resulting in more confusion and misunderstanding and lots of spoiled dogs. Traditionalists or not, we don't see a taboo on corrections as a service to dogs or their owners.

For those who are interested in investigating clicker training, we recommend getting beyond the hype over ethics and the philosophical issue of whether it is more loving to train your dog this way. Why not simply try to gauge if this method would be something appropriate for your dog? Generally, dogs with softer, more sensitive temperaments respond well to clicker training. If you're inclined to give it a try, try to attend a workshop where you can see things firsthand. Then you can compare what you've seen with how dogs do in a more traditional class. If you decide to pursue clicker training, there are plenty of videos and books to help you.

But getting back to Pavlov: we have found that Pavlov's initial discoveries have been expanded with great success in dog-behavior modification. If you have questions about using sound as therapy for dogs, it is best to consult a trainer experienced in applying it. Pavlov's lectures themselves on conditioned reflex make interesting, if heavy, reading.* Playing Pavlov can be fun and effective, limited only by your ability to understand and apply the basic principles of sound conditioning.

*I. P. Pavlov, *Conditioned Reflexes,* trans. G. V. Anrep (London: Oxford University Press, 1927).

# 22

# Silence and Your Dog

*Animals are creatures that lead silence through the world of man and language and are always putting silence down in front of man. Many things that human words have upset are set at rest again by the silence of animals. Animals move through the world like a caravan of silence.*

*A whole world, that of nature and that of animals, is filled with silence. Nature and animals seem like protuberances of silence. The silence of animals and the silence of nature would not be so great and noble if it were merely a failure of language to materialize. Silence has been entrusted to the animals and to nature as something created for its own sake.*

— Max Picard, *The World of Silence*

Once, an Irish setter named Queenie was brought to us, quivering like a leaf. The harried owner wanted to leave her with us for observation. "She shakes like that all the time," she explained. "I don't know if she can take our lifestyle." This comment inaugurated a discussion of the lifestyle at the dog's home. The woman described her family as "active" and "robust"— and, she added, "very noisy." Meanwhile, her three preschoolers were in the car in the parking lot, alternately laughing, screaming, and crying. When asked if she wanted the children to be in on the consultation, the woman exclaimed, "Oh no, they're too noisy. When they're around, Queenie shakes even more!"

As we explored the family situation, it became clear that this dog had hardly any time to herself. Except for a five-hour stretch at night when she slept, the poor animal lived under a constant barrage of noise and racket. Orders and requests in this family, whether to the dog or to one another, were screamed or shouted. The television was the nerve center of the home and was on almost twenty-four hours a day, even if no one was watching it. When the family went out somewhere, the TV and radio were left blaring because they were afraid Queenie might become lonely and launch on a spree of destruction. Furthermore, the home was situated on a busy highway, which lent its own noise to what the family produced.

However, Queenie's shaking had started only when she was six months old. The family had purchased the dog when she was two months old, and the animal was fine for about four months. But once the shakes started, they continued, and now Queenie was almost two. In all her time with these people, she enjoyed few moments alone (even defecation was while on a leash).

After two days in our relatively quiet surroundings, she stopped shaking. The problem did not appear to be genetic, nor did it seem to stem from any kind of high-strung nervousness. Trained in a park obedience program, she responded to normally spoken commands and even to whispered instructions. When with us, she seemed to enjoy herself thoroughly. She entertained herself by tossing balls, by sprinting, and by relating to her canine neighbors in our kennels. When the owners came to pick her up, they could not believe the transformation.

After suggesting some diet changes and a vet check, we then launched into a discussion of home conditions and the way the dog was treated. We emphasized daily obedience exercises, daily play sessions, and, most important, some quiet time alone for Queenie. As a result of reexamining the whole situation, the owners stopped blaming the dog for not being able to "take it" and began to consider their own behavior and the general atmosphere of their home. The children (this time included in the discussion) were especially sensitive and concerned. One boy, five, said he often had headaches all day. Pointing to their pet, he said, "Maybe Queenie does, too."

Yes, dogs need silence. They also need some time alone, even though they are basically pack-oriented animals. It's up to us to provide them with some quiet time. There is no need to feel guilty about training your dog to be alone for a certain period of time each day, as long as it is not excessive. Even when we are with them, a certain silence can be extremely beneficial to the relationship. Though most obedience training should be highly animated and lively, full of peppy talk and encouragement, it's a good idea to have an occasional quiet session, whispering the commands, moving around lightly, perhaps even conducting the whole session in a wooded area or park. In fact, you can extend this idea to your longer walks. Don't undervalue the beneficial effects of a quiet walk in the woods. Even if your dog is not off-leash-trained, use a flexi-lead* to make things easier on yourself and your dog. Many an owner has commented how such a walk seems to renew the dog and calm him down.

Eye contact works with silence as another essential element in any good canine-human relationship. It's a good practice to stop once a day, get your dog's attention, and simply look at him. Don't stare, however. Simply stop, look, and smile while remaining silent. In a word, "insee," as Rilke says. Then end the moment with an affectionate pat or friendly word. Direct stares, hard and penetrating, can be interpreted as threatening, so avoid them. The kind of silent eye

*Quiet walks with your dog are calming and help keep both of you connected with the natural world.*

*See pages 219–20.

contact we are concerned with here is gentle and sustained, a real exchange between animal and human; therefore, its entire "tonality" is one of peace and quiet.

There is a myth that all obedience training should be done in noisy environments that simulate "real life." Yet in nature films on wild dogs and wolves we can see the real periods of silence enjoyed by the pack. Distractions and other noise do serve to teach proper retention of commands. But especially at the beginning of formal training, you must plan silent sessions with your dog. Silence is such a rarity that we must plan for it, or it will not come our way. So figure it into your life, and into your dog's. Everyone has at some time experienced a moment of silent communication with his or her dog. It's important to cultivate these moments as you grow in your relationship with your pet.

Some city dwellers might well ask, "But where can I go for silence every day?" With some creative thinking, there are solutions. One urbanite takes his obedience-trained mastiff to the local branch library. He explained his situation to the head librarian and asked for

*A good dog can be a part of your study and quiet time, even in a library.*

a trial run. Another woman in a crowded suburban area stops at a local church for ten minutes every day and puts her golden retriever on a down-stay between the pews. Still another owner closes his windows to highway noise, draws the drapes, lights a candle, and does yoga while his dog lies nearby.

Some of us do a similar practice here at the monastery during our meditation periods. When a dog knows how to hold a down-stay for an extended period of time, she can be included in such intimate, silent activities. They provide the dog with a context of silence, too, one in which the dog settles down and becomes relaxed and comfortable as she does nothing. The gist is obvious; if there are already quiet reflective moments in your life, it is just a matter of letting your dog in on them. If, on the other hand, you never have any quiet time alone to retreat and refocus yourself, perhaps you should think of ways to incorporate some such moments into your own life and share them with your dog.

*A guide dog or a well-behaved pet might be able to attend church with you.*

# 23

# Radio Training

One piece of equipment in our training collection that looks out of place when lined up with training collars and leashes is the radio. You can use a radio to help train your dog. All you need to do is tune in to a station.

Playing the radio for a litter of puppies is an old breeder socialization technique. We play the radio for our puppies here at New Skete. Since most of our puppies go into a family setting, we feel that it is important for them to hear a variety of voices — men's, women's, and children's. National Public Radio's *All Things Considered* has been heard by many of our pups. During the day, they may listen to talk shows, children's shows, and many styles of music. Call-in talk shows offer a wide and constantly changing variety of voices. But stay away from "controversial" shows that feature people arguing.

Some say that rock is the best socialization music for pups, and others swear by classical music. The positive effects of Beethoven on dairy cow milk production is well known by farmers. The "Mozart effect" on human beings has been well publicized. In the movie *One Flew Over the Cuckoo's Nest* psychiatric patients were lulled into inertia by placid music that was played all the time. As far as we know, no specific studies have been done on the effects of different types of music on dogs, but it is our experience that music can be of value. But *pace* those who feel rock and heavy metal music are good socializing devices; we advise being careful not to bombard dogs, young or old, constantly with loud, assaulting noise. But do use the radio creatively, especially for problems.

The radio can help a dog spend long periods of time alone. For example, Clancy, a two-year-old Irish setter, had difficulty staying alone while his owner, a young office worker, was at work. He barked incessantly. The desperate woman brought the dog in for training after her neighbors threatened to have her evicted if the barking did not stop. Clancy began obedience training right away, mastering heel, sit, stay, and come very quickly. His owner remarked during one session how attentive Clancy was to the sound of popular records played on the stereo. "The first thing I do when I get home is put on a stack of records to help me unwind. Clancy loves it. He sits right next to the speakers and listens, almost like the famous RCA Victor dog."

This tip helped us figure out an additional tactic to stop the barking. We suggested that the owner begin playing the radio half an hour before leaving home and to leave the radio on all day, tuned to a station featuring the kind of music she usually played. She should continue to play the radio or stereo when she returned home. Greetings and departures were to be kept low-key. Combined with obedience training, the radio seemed to help the setter keep quiet. The radio masked traffic and street noises, especially the sound of children and other dogs playing in a nearby school yard, which may have helped set the dog off on a barking spree.

Looking back on the Clancy case, we are mindful that, like us, many breeders and boarding-kennel proprietors use piped-in music because they feel it quiets barkers and howlers. If you consider "radio therapy" for your dog, don't use it as the only technique to solve barking or other problems. It needs to be combined with an effective program of obedience training. And don't forget that too much noise can backfire. Remember the story of Queenie in the last chapter!

# 24

# Massage for Dogs

*When you sit down to work with a creature, simply focus on what you're doing instead of trying for a connection. Seeking too hard or waiting for some kind of heightened contact will not only block your own experience but will also make your animal nervous and self-conscious.*

— LINDA TELLINGTON-JONES, *The Tellington TTouch*

Not every culture that allows domestic pets teaches its members ways to relate physically with those animals. In some countries dogs live a dog's life and are rarely held or petted. We've noticed that some German dogs we import do not seem to like our "American" way of petting. After investigation through our German contacts, we learned that Germans have a slightly different approach to their dogs. They pet and stroke them in a different way and in different places than do many Americans.

In American culture, petting a dog is very important. Most people tend to pet dogs around the head and shoulders and stop there. Others literally trounce their dogs, pounding their sides and ruffling their fur.

Sometimes there is little method to the physical display. The dog is expected to take it whether or not it is the kind of physical affection he enjoys. Few dog owners stop to read their dog's needs and desires. A dog owner may find that the dog does not enjoy being petted — if by "petting" we mean rough jostling or pounding. Instead, like many humans, they greatly enjoy a more extended type of body contact — a kind of massage.

Be this as it may, massage is not simply a New Age technique meant to appeal to those looking for a new gimmick. Setting aside for a moment the beneficial effects of massage in terms of relaxation, there are other crucial reasons for conditioning your dog to accept your gentle touch. For example, getting your dog accustomed early to having her whole body touched and handled makes it safe and easy for veterinarians and groomers to work with her. This can head off the need for muzzles and anesthesia during office visits and creates a much more positive attitude in your dog when visiting these professionals. Related to this is the probability of your dog's becoming injured at some point in her life. Dogs who are in pain can often bite reflexively when they are manipulated, and this possibility is lessened considerably if they are used to being handled (massaged) and calmed. By calming the dog, you can more easily examine the wound and determine the best course to follow.

Regularly massaging your dog also keeps you in touch with his overall state of health. It allows you, without having to wrestle with him, to check for tumors, soreness, swelling, and fleas at the same time you are deepening the bond with your dog. This is an important aspect of responsible ownership that people often don't consider.

Finally, pet massage can be a beneficial relaxation technique for *both* you and your dog. Dogs experience stress just as humans do, and the sensation of a gentle hand helps relieve this tension as well as a variety of aches and pains. It also sends a powerful message of care and concern to your dog. At the same time, you will discover that this massage helps you relax, too. By relaxing and breathing deeply as you massage your dog, you'll find your own stresses and tensions starting to dissipate.

The first principle of dog massage is to stop thinking of your dog solely from the shoulders up. Contact can be made with almost any part of the dog's body if it is done with sensitivity. Skilled veterinarians know this from treating unapproachable patients. They often have to devise creative ways of lifting the animal up onto an examination table, or treating injuries all over a pet's body.

To begin dog massage, make a list of all the areas where the dog likes body contact. If you are the dog's owner, you know. If you are not, ask the owner. Then list the areas where the dog dislikes being touched. Begin your first massage with the areas on your first list, but

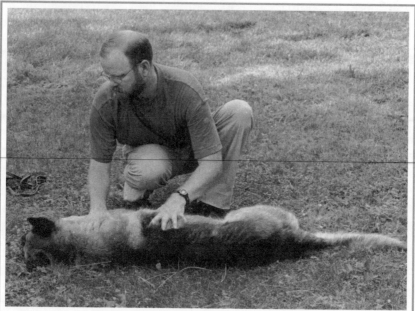

*With your dog in a relaxed down-stay, slowly begin massaging his neck and side.*

include one area on the second. Gradually include more "forbidden" areas in subsequent massages. A good place to massage your dog is on the floor. A carpeted area is best. The immediate inclination of many dogs is to play. If the dog wants to play, let her play. Don't make massage a businesslike experience. Use your voice and hands to calm her down. You might begin with the dog in the sitting position. It's best to begin on the head, gently massaging the eyelids, muzzle, and nose. Always keep one hand in contact with the dog during the entire massage.

From the head area, work down the neck, to the chest and pectoral muscle. Some dogs automatically offer a paw. Take hold of it, but gently place it down if the dog seems to be losing balance. Remember, there is no absolute right way to give a massage — as long as the dog is enjoying it.

Choose a leg and work up and down on it very gently. If your dog decides to lie down, you have better access to his rear legs. Try to avoid forcing the dog down. If your dog knows the command for down, you can use it in massage work, but don't force the issue. Make your

*Gently work each of the legs and the back and chest. Avoid slapping, pinching, or pulling.*

*Working on the neck muscles and the head and face can be enjoyable for both of you.*

strokes long and firm or short and circular. Try to distinguish massage from regular petting. The massage should be more extended and pliable in its movement than regular petting. Avoid all slapping, pinching, and pulling motions. These will break the mood of the massage.

Many dogs communicate quite clearly what they like and dislike. For the owner who has never had such extended contact with his or her pet, it may take a while to feel comfortable. Try not to be embarrassed and timid by focusing on your breathing and a calming touch simultaneously. If you are nervous during the massage, your dog can sense it and tighten up.

You may find that some dogs are initially too restless to stay for any extended massage. This is fine. Simply start with a very brief massage and then praise her warmly, telling the dog how good she's been. You can build on this foundation by gradually extending the length of a massage. Also, another way of helping settle down an active dog may be to exercise him vigorously beforehand.

There are many benefits to dog massage. Linda Tellington-Jones, a well-known animal consultant and trainer, has developed an entire training method based on touch, which can be applied to a number of different species, including dogs. Building an entire method on touch and massage may seem to be taking things a bit far, yet her work makes for interesting reading, and the general thrust of her argument bears serious consideration. Massage is also a veterinary technique used to hasten rehabilitation following fractures and luxations and to restore muscle tone. It can give you a new appreciation of canine anatomy. For the busy dog owner, it is a way of disciplining oneself to make contact with the pet. It is a welcome break for dogs in obedience training. For show dogs, massage is an excellent calming procedure before entering the ring, where a delicate balance between relaxation and animation must be maintained if the dogs are to look their best. Acupuncture and chiropractic are also used therapeutically on dogs, just as they are on humans.

A technique for calming down a stressed dog is to place your open palm over the dog's groin area and hold it there. On females, this is just below the stomach; on males, just in front of the genital area. The groin area is a traditional greeting place between fellow canines. Dogs often tell each other "it's all right" by nudging this area. When you gently place your hand there, it has the same effect on the dog as an arm around the shoulder or a handshake can have for worried humans. Try this technique if your dog becomes agitated while at the veterinarian, when you need to groom him or cut his toenails, or anytime the dog is under stress.

We often suggest massage in the case of an overworked or overstressed owner, or in neglectful dog-owner relationship. Children are often great at dog massage, once they understand what it is about. It is a good substitute for unhealthy kinds of contact between kids and dogs, such as ear pulling, chasing, and tail yanking. If you have a child, monitor several sessions with your child so that both dog and child get used to it. The dog owner who finds it difficult to reach his pet, especially on the verbal level, can often calm the animal down with massage enough so that she can learn and retain commands. We have had success with hyperkinetic dogs by including a five- or ten-minute massage session before commencing regular obedience work.

# 25

# The Round-Robin Recall

Dogs with recall problems can be helped with a simple game the whole family, a group of friends, or a class can play. You need a twenty-foot rope or lead with a light weight attached to one end. Your dog should wear a training collar and should already be trained to the come, sit, and stay level before attempting round-robin work.

Begin by forming a small circle of four or five people (family members first). The object of the exercise is to call the dog, have the dog sit in front of you, praise him, and toss the rope to the next person in the circle. This person then calls the dog and repeats the process. The tossed rope is to ensure a prompt recall by the dog. If he does not come after being called twice, give the lead a quick pop and call the dog, praising him warmly as he trots toward you. As the dog comes to each person in the circle, that person should have the dog sit and then give lavish praise again. A treat can be given in conjunction with the praise as soon as the dog sits.

All participants should make this a lively, fun session. It should not be a formal, precise lesson, but an experience in animation and praise. Keep the dog happy by giving encouragement. As soon as the dog is called and even looks at the caller, that person should start giving the dog encouragement. But if the dog does not come, or begins to go the other way, bring the dog in with a quick pop on the lead, calling him at the same time.

Be sure to have a session like this twice a week if you are having problems with the come exercise. Gradually extend the circle physically and psychologically. Broaden the circle so that the full length of the lead is between each participant. At first the circle should be made up of the dog's immediate family-pack members, but including strangers after a few weeks of practice can often improve the behavior of aggressive dogs.

## How You Can Use This Exercise

Dogs are often so attached to one member of the family that they ignore or refuse to obey others. Regular round-robin sessions help focus them on other family members, particularly if those members also take on responsibilities, such as feeding and walking the dog (if only for a limited amount of time). This helps promote a proper leadership role in each family member. In such situations, it might be best to exclude the dog's favorite from the round-robin circle at first. Have that person stay out of sight, thus forcing the dog to relate with other persons who are less desired.

Don't take it personally if the dog doesn't come to you immediately, or if she comes slowly and reluctantly. That is why you have the lead and use your voice. Be animated and encourage the dog, and give the lead a quick tug if necessary. Praise the dog when the recall is complete. Don't let a session go longer than ten or twenty minutes, and end it on a playful note. We've seen remarkable improvement in recall and aggression problems after using the round-robin technique several times a week.

Clients also report other positive changes after a few sessions. Dogs for whom round-robin sessions have been prescribed begin to relate in new ways to family members they had formerly ignored. After a few initial sessions, many dogs begin to whirl around the circle very quickly, barely stopping for praise from each handler, eager to get to the next. As one client reported after three round-robins with a former fear-biter, "Jake zips around the circle like crazy. He loves it and it's given him a new lease on life. Our family really enjoys it, and we have a waiting list of area kids who want a turn in the circle."

# 26

# Keeping Fit for Life

To provide a city dog with more vigorous exercise in a congested environment, urban dog owners might consider roadworking their dogs as an alternative to daily strolls. In fact, it can be useful in any environment. Daily exercise is a requirement for all dogs, and owners need to look for creative ways of providing it. Swimming is one possibility; however, roadworking is another. Roadworking is definitely not a matter of hooking up a dog with a rope or chain to a car bumper and then speeding off down the road. A dog should never be tethered to a motor vehicle, even when parked. However, there are a number of safe ways owners can provide much-needed exercise for their dogs that make use of a bike or vehicle. Assuming that your dog has a proper foundation in obedience training, that you take proper precautions and go slow, and that you work in a quiet, level area, roadwork can be quite beneficial.

Dogs needing better weight control and ligament tightening can benefit from roadwork. Certain very active breeds such as Australian shepherds and border collies need a lot more exercise than simply a walk or a short Frisbee session, and they respond well to roadwork. In fact, most dogs could use a lot more exercise than owners provide them. You should start walking your dog with a fifteen-foot lead, preferably on dirt, sand, or grass. Do not allow rowdiness in the walking phase. Then you can phase in jogging, with your dog trotting ahead.

Next try a bicycle, if you wish, with your dog running alongside. Actually, we prefer this form of roadwork because it is safe while at the same time healthy for both dog and owner. Also, it does not require

*Well-made devices for bicycles can keep the dog safely in position for exercise.*

more than one person, the way roadwork with a car does. There are flexible devices that you can attach to your bike and your dog's collar. Practice first in an empty parking lot or playground, making right and left turns, allowing your dog to get accustomed to moving with the bike. It's a good idea to give a specific command with the turn, such as "left turn," so that the dog gets used to the protocol. Make sure that you let the dog determine the speed. Never forge out ahead of your dog, and never, ever drag the dog.

Another possibility mentioned by contemporary trainer Cis Frankel in her book *Urban Dog** is Rollerblading with a dog. Obviously, this requires a real proficiency on your part as well as reliable automatic stops and sits at street corners from your dog, but for an adept owner whose dog is well trained, it can be a lot of fun and a great workout for your dog. As with any sort of roadwork, always be careful not to overdo things.

Finally, if your walking, jogging, and bicycling are smooth, you can begin to work your dog off a car, but it is vital to follow safety precautions. When roadworking, always use wide, flat nylon collars,

*Cis Frankel, *Urban Dog* (Minocqua, Wis.: Willow Creek Press, 2000), pp. 152–53.

*Regular roadwork for exercise can be done safely and enjoyably. Keep the lead loose and the dog away from the car.*

as they prevent any sort of unintended constriction of breathing. Choose secluded roads with good shoulders. Never try to handle roadwork in a vehicle alone — this can be dangerous since it is possible for the dog to get too close to the car. Get a partner to drive while you control the leash. If necessary, you can even string the leash through a lengthy plastic tube to keep the dog sufficiently away from the car. Make sure the driver does not go too fast (don't go more than five mph). The dog should be moving at a very comfortable trot. Never work a dog more than two miles, no matter what shape he is in. City dog owners can roadwork their pets in local parks or drive out of town to unpopulated areas. Elderly people who are not physically spry but who like large-breed dogs can exercise their animals with a roadwork program. Make sure the leash is well out from under the car, and never allow the dog enough leeway to go under a tire.

Make roadwork a highly ritualized and enjoyable event. While roadworking, it is essential to shout encouragement to your dog constantly. It might mean going every day to an area where you can yell and scream without disturbing anyone else. We prefer the early-morning hours when the pavement is cool, the sun just peeping up, and the air crisp and fresh. It is an excellent way for a dog to start the day. Avoid working in the hot sun. Because of their hair, dogs do not cool themselves very effectively. When they exercise too vigorously in hot weather, they become susceptible to overheating and heatstroke. If you are able to work close to home, first drive to your starting point with your dog loaded in the car, then begin the roadwork, aiming your dog for those highly desirable goals: home, food, water. This will greatly improve the dog's attitude and drive.

## Benefits of Roadwork

Some people think that roadwork is only for show dogs with spongy constitutions who need a crash program of heavy exercise in order to meet a show deadline. We mentioned the benefits of roadwork especially for the urban pet. But roadwork has many benefits for all dogs. Bitches who have recently whelped a litter profit from this form of exercise, which restores their muscle tone after the strenuous nursing period. Serious obedience fanciers will find that roadwork greatly improves trainer-dog relationships and is a welcome break from more precise training routines. Pet owners with hyperactive or seemingly "uncontrollable" canines have much to gain from it, too. The whole roadwork ritual removes both dog and owner from home stresses (telephone, chores, children) and builds good rapport. It vents energy and frustration for the high-energy dog and the high-energy owner. We recommend it as therapy for pet owners with dog-management problems.

*As an alternative to roadworking, swimming in a stream or at the beach is both great exercise and a time for play. Remember to rinse off salt water.*

# 27

# Avoid Canine Incarceration!

If your dog spends any length of time in a yard, you might be interested in ideas that can help you make his stay there a pleasure, not an imprisonment. It is amazing how many dog owners have described their dog's play area as a gravel enclosure with nothing, absolutely nothing, inside, except the dog. Boredom and ennui are one of the worst aspects of modern dog life. Chewing, digging, nuisance barking, poor appetite, and stool eating are usually in some way connected with boredom. These problems often occur in dogs who have become "kennelized" by such uncreative play areas.

Before the professional dog people get up in arms, let us note that we are not against the use of kennels or enclosed areas. They are necessary. But they can be built with imagination. It doesn't matter whether you have one dog or many, a yard can be well thought out. First, the shape. For dogs who are alone, rectangular pens encourage dogs to walk and strut about; square kennels encourage them to lie down and do nothing. Unfortunately, the rectangular dog run may also encourage fence charging in some dogs and endless barking, which is linked to the tension of fence running. If your dog is prone to fence running, change the yard pen to a square design, and the running may stop. For large working breeds, we usually suggest square pens of at least twelve feet by twelve feet, although a smaller square can be used if the dog is not placed in one for too long.

It is important to note that in discussing play-area kennels, we are referring to a pen in which the dog spends a sizable amount of time, not simply defecation runs. Enclosures constructed solely for defecation can be quite small. More than eight feet in any direction is rarely

*Teaching a pup how to use the playground helps cut down boredom and instills confidence.*

*Gently leading the dog or pup on new surfaces helps teach her to trust your lead.*

needed for any dog's elimination needs. We suggest this type of pen so that the owner can more effectively clean up after the dog instead of letting the dog squat where he will or eliminate on the curb. Even curbed stools are health hazards, although they beat a mess deposited in the middle of the street, sidewalk, or a neighbor's yard.

Rectangular or square, your kennel yard should be a fun place. A simple obstacle course of old tires, a curved board set in concrete at each end for a nifty bridge, a tunnel, and scratch posts can occupy many dogs for hours.* Provide dog-size toys. Owners of toy breeds might be able to substitute children's toys, but larger breeds need larger toys. You can construct your own, as we often do, with crunched-in bleach bottles, PVC pipe, blocks of four-by-fours (untreated), old broom handles, leather scraps, and bells. Just make sure that all sharp edges are sanded down and that no toy is so small that your dog can swallow it.

Another possibility is to provide toys that are designed to give your dog something to do, for example, plastic cubes that can be filled with kibble, which the dog has to work at for it to be dispensed, one kibble at a time. Also, there are durable plastic balls that have a recessed screw which allows sand, water, or gravel to be put inside. As the dog plays with the ball, the sound is changed as well as the way it rolls, piquing the dog's interest. They are virtually indestructible and, once he is taught how to play with them, can occupy your dog for hours.

Dogs love any hanging object, especially if it makes some kind of noise. Suspend toys and leather scraps (ask for them at a leather shop) from strong ropes. Always use single-strand ropes for hanging toys, and never arrange a hanging toy so low or in such a way that a hanging accident might occur. Rope toys suspended by springs make the toy snap back when the dog lets go, so the dog can play fetch alone. Hanging toys with bells attached keep pets fascinated, but make sure the noise does not bother neighbors. However, most neighbors prefer the sound of bell chimes to incessant barking.

Try to avoid barrier frustration (see chapter 43 on aggression) by screening the dog's area so she can't see busy city street traffic or passersby. If you use cyclone fencing, standard green slats that fit

---

*"A Puppy Obstacle Course" by Brother Job Evans in the May 1977 issue of *Off Lead* training magazine is still relevant.

between the chain link are available. Shrubbery can be effectively arranged to block disturbing views that encourage barking and fence running. Within the pen, a ditch or a wall, or even a row of flowers, can often keep the dog away from the fence and in the center of the pen. Gravel is the best all-around footing, but dogs sometimes ingest small pellets of pea stone, which can be dangerous. Concrete, grass, macadam (provided there is enough shade in the summer), and even wood may be feasible. The exercise pens we provide for our dogs are a combination of cyclone fencing and wood-plank floors, with wooden walls to prevent dogs in adjacent pens from seeing each other. The hemlock floors (each slat is spaced one-eighth of an inch from the next) have proved durable in our upstate New York climate and provide excellent drainage in wet weather and dry quickly. Floors made of concrete lead to paw sores and cut pads, so avoid prolonged contact with it. Concrete is better for a small run where the dog "does his business." Pick up and disinfect daily. Grass inevitably wears out, but it is aesthetically pleasing and comfortable for the dog. We find a grass pen with gravel along the sides of the fencing to be the best. The gravel prevents the grass from being totally destroyed if the pen is a very large fenced-in area, say twenty feet by twenty feet or more.

Each day introduce a different item into the pen, especially if you leave your dog alone on a daily basis. Favorite toys should always be included, but rotate others to spice up the routine. Rotate water and food to different locations in the pen. When you prune your trees, throw the limbs into the pen for the dog to play with. A large cardboard box enthralls any dog. Sure, the box gets ripped up and the tree limbs get shredded and you have to clean up. But it's worth it in the long run. Fresh fruit and vegetables can be used as vitamin-packed toys. Toss a few into the pen every so often. Bones, if offered, should be large marrow factory bones that have been boiled for five minutes. A dog left alone with a small bone all day can reduce it to such a small size as to risk lodging it in his mouth or throat. Make sure bones are large and solid, but not cooked.

The doghouse need not be elaborate but should provide protection from the elements. In summer, repaint dark-colored houses white to reflect sunlight and heat. Give the dog some options for shade in addition to the house. Trees are best, or a planter of bushes, pruned of their lower branches and with trunks wrapped in tree tape to prevent the dog from destroying or scratching them.

Finally, remember what we said in our discussion of canine loneliness: consider providing your dog with the ultimate diversion and plaything, another dog. Owners with chronic diggers, chewers, barkers, squealers, and house wreckers have often found that the introduction of a second compatible pet (it could also be a cat) can reduce the severity of this behavior. You have to be sure that the new animal gets off to a right start and does not mimic the bad behavior of the first pet. For dogs who must stay in play yards for long periods, a companion can change frustration and boredom to pleasure and play.

Break up monotony whenever and however you can. Use these techniques to eliminate undesirable pen behavior and encourage proper exercise and play. Give some thought to what your dog does during the day, watch for his special interest in toys and devices, and capitalize on this interest. A diversified and creative environment is essential not just for human beings, but for dogs as well.

**28**

# Children and Dogs

We have long been aware that dogs can have a profound and positive influence on the emotional and spiritual development of children. Throughout the history of our community we have had the chance to observe how children visiting the monastery interact with their own dogs as well as with our German shepherds. A mutually beneficial effect has always been palpably evident to us: introduced and nurtured correctly, dogs enrich the lives of kids just as kids enrich the lives of dogs.

In fact, this perception is confirmed by modern research into the dog-human bond.* Studies have shown that children who help care for dogs often benefit in peripheral ways that make them more adept at social relationships, making friends more easily and generally developing a greater level of self-confidence. They are also generally more benevolent toward others. These children's relationships with dogs are profoundly humanizing. Dogs are not meant to replace friends and peers, but they can help relieve a child's sense of isolation, making the movement toward others less awkward. Significantly, this is even more the case for children with various disabilities. Studies have documented how service dogs who work with disabled children help them normalize their relationships with their peers. When compared with that of disabled children who do not have a service dog, their social contact is much greater.**

*See Lynette A. Hart, "Dogs As Human Companions: A Review of the Relationship," in *The Domestic Dog: Its Evolution, Behavior and Interactions with People*, ed. James Serpell (New York: Cambridge University Press, 1995), pp. 166–68.

**Ibid.

Author-trainer Mordecai Siegel is very perceptive in identifying the important role dogs can have in the development of children, articulating forcefully what many parents intuitively sense:

> Of all that children derive from their pets, including understanding birth, death, growth and relationships, one of the most worthwhile lessons is leadership. I don't mean leadership in the sense of high-school graduation speeches or militaristic virtues, but rather in the vital areas of independence, self-sufficiency and competent self-management. If leadership is understood as the willingness to assume responsibility, to make decisions, to risk failure, then a child living with a pet is, indeed, involved with leadership. Dogs are capable of bringing out the best in a child or helping parents create values that are more learned than inherited.*

To include dogs as a vital component to the development and maturation of children, it always helps to plan ahead and clearly think through their introduction into the family. When you simply say that a child "needs a dog" and leave it at that, or you get a pup simply in order to provide a playmate "to get the kid out of my hair," you might do well to think again. Though we applaud children's actively participating in the care and training of a dog *as their age allows,* we've learned along with many experienced parents and trainers that it is not reasonable or appropriate to expect an eight-year-old to take care of the dog. A child of that age is still not capable of this multilayered responsibility, especially when required to do so alone. Lassie and Timmy as inseparable companions may be fine for Hollywood, but Hollywood is not the real world. To burden a child with full responsibility for another living being as the price for its companionship is both unwise and unfair. Over the years we have listened to too many parents complain that they originally obtained their dog "to help teach their child responsibility," but in the end they were the ones who had to assume primary care of the dog.

In particular, many a mother has inherited the task of feeding, caring for, and walking a pet who was originally supposed to be the

---

*Mordecai Siegel, *A Dog for the Kids* (Boston: Little, Brown, 1984), pp. 7, 8.

responsibility of the child. It's no surprise that much of the pet advice in our culture is offered in women's magazines. Women seem most often to get stuck managing the everyday lives of their kids in our culture, so they also end up with the children's dogs as part of the bargain. Smart parents see this coming and avoid walking into such a situation unprepared. Unless the mother (or increasingly, the house-husband) can accept the full role of being primary caregiver to the dog while the child is younger than twelve years of age, the family should hold off on the decision to get a dog. There may be an instance of a younger child having the maturity and interest necessary to assume this primary role, but it would be exceptional. Most children work best with a pup when they are in a supporting role, the oversight resting with an adult.

## Foundational Principles

Once a family of six visited New Skete. They were interested in raising a German shepherd and admired the breed, but they expressed concern over incidents they had heard concerning children and German shepherds. At that discussion we were surrounded by several grown dogs we were evaluating together. The father of the family asked the monk who was showing them the dogs, "What is the point of a good breeding program?" Just then, a three-year-old toddler waddled over to one of the dogs, grabbed her tail, and gave it a good yank. The surprised dog moved a few inches, turned, and licked the child's face. The monk turned to the father of the family. "That," he said, "is the point of a good breeding program."

Still, it is folly to presume that dogs will always behave with such equanimity, particularly with respect to children. Dogs tend to view them as peers, even subordinate members of the pack. If you ever observe how dogs establish hierarchy in a pack, you will see that dogs are not genteel with one another. When children are not educated correctly in how to behave with a dog and supervised accordingly, dangerous displays of canine dominance can easily follow. Always be on the lookout for such instances. Although completely "natural" from the dog's point of view, they are totally inappropriate in a family situation and, if acted upon, often result in your having to get rid of the dog. Parents have a crucial responsibility to socialize both dogs and children with one another. Dogs need to be trained how to act around children, and vice versa. The latter includes more than simply teaching children how to approach a dog: it is an entire process of socializing, one that gradually teaches a child how to respect a dog as another living creature.

Such an understanding is imperative because children in our culture are exposed to a high degree of anthropomorphic conditioning through television and books. They are constantly exposed to animal figures who act like humans. We have already mentioned Lassie, but there is a long list: Goofy wears human clothes, drives a car, has a girlfriend. Mickey and Minnie Mouse set up housekeeping long ago. Rin-Tin-Tin finds the robbers, saves the family from a burning home, and attacks all the right people. Fairy tales abound with animals that have human traits. Pet-food commercials are filled with animals that talk, dance, sing, open beer bottles, and generally behave like humans. To a child, the dog is a buddy, another child. Children think of dogs as other people.

An excellent illustrated book by Maurice Sendak and Matthew Margolis, *Some Swell Pup*, is a children's story that attempts to portray realistically what is involved in purchasing and raising a puppy. This is a good book to read with your children if you plan to bring a puppy into the home. From early childhood, try to balance anthropomorphic thinking in your child by providing him or her with realistic stories about dogs and other animals. If your family has not as yet acquired a dog, make a visit with your child to a pound or shelter to expose him or her to interaction with a dog, the variety of breeds, and to the problems of the pet population. If there is an obedience class in your area, why not take your child to it and watch together from the sidelines?

Once a family obtains a dog, parents have the responsibility of teaching their children never to tease her. Unfortunately, teasing is an occupation many children thrive on: it relieves boredom at the same time as it entertains. Children are often unaware of the effects of poking a dog, pulling tails and ears, and running and screaming around a dog. They can easily miss signs the dog is giving that an aggressive response is in the offing. Before you know it, an unfortunate experience can occur. This is your lookout as a parent.

Not surprisingly, the most important thing you can do to teach your children how to act toward a dog is setting a good example. Over the years in our work, we have seen that children tend to mimic the behavior of their parents toward the dog. If you treat your dog with consideration and respect, your child will see and tend to imitate that attitude. Children have fertile imaginations. By encouraging them to look at things from the dog's perspective, parents can help guide their kids into asking themselves, "Would I want to be treated in such a way?" Additionally, including children in the ordinary chores involved with caring for the dog helps cement the bond of friendship in a way that makes it more difficult for the child to tease and abuse the dog.

What about if you don't have a dog yet? A time-honored technique to help younger children learn how to interact with a puppy is to have them practice with a dog doll first (i.e., teaching them how to pet). We prefer encouraging parents to expose their children beforehand to real-life situations with trustworthy dogs and pups. These can often be arranged with friends and neighbors and have the virtue of preparing children for life with a pet by providing them with hands-on experience. This is a good idea even if you don't want a dog in the near future, for your child will inevitably encounter dogs while playing at friends' houses. Over the years we have noticed that some parents can be overly protective of their children, screening them from contact with dogs for all sorts of reasons. Yet children not exposed to dogs early in their lives in a controlled and natural way can very easily become fearful or act in ways that invite aggressive responses from a dog. We advise parents in general to teach their children proper manners around a dog whether or not their family has a dog. Here are some basic guidelines we have found helpful:

1. Never approach a dog while he is eating. Dogs instinctively protect their food, and little children who approach them at this time may provoke an aggressive response. It is also a good idea for you as an adult to desensitize your dog to protective behavior around food. This means using a progressive series of behavior-modification exercises grounded in feeding him out of your hand. But it is also wise to think preventatively. Especially when there are small children around, don't invite trouble. Have the dog eat his meal in his den or in a quiet spot.

2. Never approach a strange dog who is not on leash with her owner. Despite the fact that some dogs may appear to be friendly at first, dogs are capable of a quick, aggressive response if a child

*It is important to introduce your dog to children in a deliberate, nonthreatening way.*

suddenly acts unpredictably. Instead, calmly instruct your child to avoid the dog. Similarly, children need to be taught that they should never enter a fenced-in yard where a dog is loose. Dogs are territorial creatures and often act with hostility when a child enters their space.

**3.** Never approach or disturb a dog who is sleeping. There are incidents galore on record of dogs who, when suddenly awakened, have nipped an owner or child. If the dog must be roused, teach your child to call his name from a distance while clapping her hands. This alerts the dog well in advance and gives him the chance to respond evenly.

**4.** Finally, never come up from behind a dog suddenly. This can have serious consequences no matter how stable the dog is. Dogs generally do not enjoy surprises, so make sure that your child has learned always to alert the dog of her presence before she approaches.

We believe that the best way to teach your child how to meet a dog for the first time is to have the owner bring the dog up to the child on leash, making him sit in front of the child. Doing so sends the dog an implicit message about the owner's authority and confidence and allows the owner to better control the introduction. At that point, have the child put her hands out for the dog to sniff, making sure that her fingers are turned inward, back of the hand forward, toward the dog. This is because the palm of the hand has many nerve endings that, in the event of an accidental bite, could be damaged easily. The back of the hand is much less vulnerable. After the dog shows signs of friendliness (wagging his tail and licking, etc.), the child can then be allowed to pet the dog more affectionately.

On the other hand, it is important that your children know how to act if and when they ever accidentally come in contact with a dog who is acting aggressive. Dogs trained in guard work learn to attack a raised hand, so if your children are accosted by a strange dog, caution them to stand still without throwing their arms up and, then, once the dog retreats a bit, to walk away slowly. Above all, they should not scream or run away, which might encourage the dog to attack. If you know of an unfriendly dog in your neighborhood, notify the owners — don't wait for an unfortunate incident to occur.

## Children and Training

Over the years a number of the children of our parishoners have helped out with kennel chores and socializing puppies. We've found that with proper supervision and monitoring, children can feed the dogs, take them for walks, help out with obedience exercises, and play with the dogs. This has been an enriching experience not only for children and pups but also for us. Children of all ages have something to give to dogs, from infancy on. For example, even an infant, if introduced properly to a dog, can provide him with a new sense of responsibility. While obviously a parent will *never* leave a dog alone with an infant, it is startling to observe such a dog take cognizance of a child's vulnerability and *share* in guardianship, lying quietly by a crib or playpen.

As the primary caregiver, of course, training is your responsibility. However, kids can be included in the training process, learning elementary commands that the dog can be trained to respond to. If you are planning to take your dog to an obedience class, find out if your children can participate along with you, or at least watch from the sidelines. Carefully including the kids in your dog's training is an ideal way to keep their interest up and to establish a context wherein both people and dogs behave in acceptable ways.

Why is this healthy context important? Dogs are not robots. They do not automatically respond to commands that people (especially children) give them, nor should they be expected to. Untutored children often bark out commands, expect the dog to comply, and then become angry and frustrated when the dog does not respond. A context of obedience needs to be established first. This means a relationship based on understanding and constant practice. Be sure to practice with your child daily, teaching him how to work with the dog properly and how to give commands the dog will respond to. Never allow the child to punish the dog. If there is ever a behavioral problem, tell your child to alert you to it first. If the child attempts to correct the dog, it is too easy for her to respond dominantly in a manner the child will be unable to control.

This advice touches on several other points. It is important that you explain your own role in disciplining the dog, what it means and why it is being done, in a manner the child understands. Children who have not been taught this often become hysterical when a dog must be restrained or disciplined, creating more havoc than the original incident. When children already understand the role of discipline, chaotic scenes can be avoided.

It is also important to explain how the children themselves should behave around the dog. Parents need to emphasize from an early age that dogs are not children and to enforce proper safe behavior around a dog. Children should not be allowed to scream or race around the pet, which can result in aggressive incidents. Tug-of-war games, sexual stimulation of the dog, or "siccing" the dog on others should be taboo. With very young children some tail yanking, ear pulling, and rough handling is inevitable, but you can minimize it if you monitor your child carefully.

These points underscore once again why it is useful to ask whether the puppies have been exposed to children when you go to buy one. Remember that the outgoing, firebrand type of pup is not necessarily the best for an active family.

# Newborn Socialization*

Over the years a common question we have been asked is what prepa-rations expectant parents can make with their dog before the arrival of their new child, to ensure a safe adjustment by the dog and his accept-ance of the child. It is always disconcerting to parents to witness their dog growling at their crying infant. Fortunately, parents can take a number of steps to minimize the risk of a tragic incident.

1. Prepare the dog. If your dog has had obedience training, make sure to sharpen up his exercises beforehand. You will be relying on them to demand a certain type of behavior of the dog in the presence of the infant. If there has been no formal training, get it. Don't pro-crastinate.

2. Well before the birth, have your dog on a down-stay as you prac-tice cradling a doll and giving it attention while a tape of a baby cry-ing is playing. This helps condition the dog to the sound of an infant. You can also practice walking your dog with a stroller — again, well before the actual birth. After the baby is born, have the father bring home a piece of clothing the baby has worn (not a dirty diaper) from the hospital in advance of the infant's arrival. This lets the dog become accustomed to the baby's scent before the actual arrival of the mother and infant.

3. When the mother and infant return home from the hospital, have the mother enter the house first to greet the dog while the father waits outside with the infant. The mother should spend some time simply being with the dog, as they will have been separated for a couple of days at least. Then the father can enter the house with the baby. Make sure the dog is on leash. Presuming that the dog is reason-ably calm, the parents can gradually allow the dog to investigate the baby, perhaps with the mother sitting on the couch cradling him.

4. Be careful not to marginalize the dog. If all the parents' atten-tion is suddenly focused entirely on the infant, the dog is likely to feel

*In the following section we are indebted to the insights of Dr. Nicholas Dodman in his book *The Dog Who Loved Too Much* (New York: Bantam, 1996), pp. 85–101, which we have incorporated with our own.

jealous and view the baby as a competitor. Work on including the dog in the new schedule with the baby. Talk to the dog when changing diapers, put him on a down-stay while nursing, or have one adult play with the dog while the other amuses the infant. In short, spend as much time as possible with both the dog and infant so that any sort of competitiveness is minimized.

5. If, in following these guidelines, you still find your dog's response untrustworthy and disconcerting, contact an animal behaviorist quickly to determine the seriousness of the situation and the best means of addressing the problem.

# Starting Off Right

# 29

# Puppy Training

Our experience in breeding and raising German shepherds for more than thirty years, helping their new owners manage the beginning stages of their relationship, and working with older dogs of all breeds and their owners has made us profoundly conscious of the possibilities in a dog-human relationship. The single most important lesson we have learned is the crucial impact of the first six months of a pup's life on his development. To put it bluntly, they are so critical that if certain social experiences are missed at precise times during this period, the negative effects on the puppy can be permanent.

For example, pups who are not socialized actively with human beings between five and twelve weeks of age are almost certain to be fearful and skittish as adults. No amount of remedial training will ever be able to overcome completely the effects of such neglect. On the other hand, puppies who are constantly exposed to a variety of social experiences (people, kids, other dogs, different surfaces, sights and sounds) during the same time frame almost always develop into happy, balanced pets. They grow to be companions capable of a wonderful relationship with a human being that ordinarily lasts the next ten to fifteen years.

This is why we have spent so much time and energy educating owners about puppy development and the vital role of socialization. If an owner gets it right at the start, so many good things are then in position to occur. Our puppy book, *The Art of Raising a Puppy*, and the second tape of our video series, *Raising Your Dog with the Monks of New Skete (In the Beginning)*, describe this process in great depth, and we encourage any of our readers who are raising (or even contemplating raising) a pup to consult these two resources. Since they are devoted entirely to puppyhood, they are far more detailed and complete than what we are able to present here in a single chapter. In what follows, we discuss the essential concepts and techniques of the book and video and introduce several new techniques we've learned more recently that can help you become more successful in raising your pup.

By bringing you into our world here at New Skete, we hope to give you a more complete picture of puppyhood. We can help you make sense out of puppy development and socialization by providing practical guidelines for establishing a healthy relationship with a pup and for becoming a benevolent leader. A reliable house-training procedure and preliminary training principles fill out the picture. The more you can understand about puppyhood, the better off you'll be in the long run.

## Boris the Chow Chow

One of the more memorable training experiences we've had over the years involved a six-month-old chow chow named Boris. The owner was at his wit's end with the dog, confiding to us when he arrived, "I wasn't sure I was going to be able to last till his six-month birthday. Thank God something can finally be done!" When we asked what he meant, he showed us several puncture wounds on his hand and arm. "But actually, that's not the worst of it," he said with a sigh. "In general he's a nice enough fellow, except when I'm trying to groom him. The real problem is that he pulls both me and my wife down the street mercilessly. Since he hasn't learned any commands yet, there's no way either of us can control him." As the attending monk crouched down to let Boris meet him, the dog backed up a few steps and started to growl. "Oh, and he doesn't like strangers," added the owner somewhat matter-of-factly. We asked him if he had taken Boris to puppy kindergarten, but he replied, "Nah, I don't think Boris would go in for that touchy-feely kind of stuff. Besides, I knew that the real training would be covered once he came here, after he was six months old. I mean, that's what you said in the letter you sent me, that he had to be six months old."

The man was confused as to what training actually is as well as when puppy training should begin. It is true that we do not accept dogs into our training program until they are six months of age, but this is because it is more important to have a pup bond solidly with his new owner before any sort of extended separation, not because the pup is incapable of starting the training process any earlier. Our training program is three to four weeks long, a length of time that would invite too much stress on a younger pup. Had the man begun to work with Boris as soon as he bought him (in a manner appropriate for Boris's age), he could have begun to shape Boris's behavior by teaching him the elementary commands in a playful, fun-oriented way. He could have also begun to establish himself as a benevolent alpha leader, encouraging Boris to respect his touch and allow grooming without snapping.

Regrettably, like so many other puppy owners, the man had acted on the myth that training cannot begin until the puppy is past the sixth month, thereby consigning owners to four months of desperate chaos before they can get their pups in for training. Fortunately, we were able to train Boris well, yet the process of formal training was much more challenging than it needed to be. The fact is, training can begin as soon as you get your pup, as long as you understand what we mean by "training" and use gentle, behaviorally sound methods to establish a foundation for obedience. Indeed, most puppies are intelligent and willing to learn, if taught correctly. As we have seen, the learning process for a puppy begins at birth, and training can begin as early as the third week. In this chapter we show how to teach your puppy standard exercises and to expose your pup to a broad range of experiences that will deepen your relationship while the puppy is still growing, before he's reached his six-month birthday.

## Kindergarten Puppy Training

Kindergarten puppy training (or KPT, as it is called) began years ago when trainers like the Pearsalls recognized its worth and started holding KPT classes. We owe these early KPT trainers a great debt for helping educate the public to the possibilities of puppy training. Researchers like Dr. J. P. Scott and Dr. J. L. Fuller laid the groundwork for KPT by demonstrating in laboratory studies the full range of puppy learning ability. Their study, *Genetics and the Social Behavior of the Dog*, is still a classic in the field, illuminating the vital relationships between genetics, early experience, and adult behavior. They demonstrated conclusively that puppies pass through four clearly identifiable stages or "critical periods" on the road to becoming a mature adult and that it is essential for certain specific experiences to happen to puppies at these times. Puppies not so exposed were found to suffer serious impediments to their personality development.* Clarence Pfaffenberger and Dr. Michael Fox later amplified these findings and, along with William Campbell and Jack and Wendy Volhard, actually developed separate puppy personality tests to help breeders and prospective owners discern the basic personality of a pup at seven weeks of age so that they could place an individual pup in the optimal situation.**

Many breeders socialize and train their own stock at a young age and encourage their clients to continue doing so, and the best literature on puppyhood and puppy training emphasizes its importance, yet the "you can't train a dog before six months" myth is still prevalent. It is true that regular leash training as described in many adult training books can be harmful to a young puppy. A pup's skeletal structure is not fully set until six months. The leash correction can be too sharp and possibly traumatic for the pup. But other, less manipulative, positively reinforced training techniques are highly effective. Formal obedience training after six months is much more difficult for a dog who has not had any prior training.

If there is a KPT class in your area, do enroll in it. The class probably meets on a weekly basis and includes between five and twenty puppies and owners. You can expect the course to cover a general introduction to puppy behavior, practice with correct procedures for socializing young puppies with other dogs and people, and an introduction to the basic exercises. KPT uses fun and play, positively reinforced techniques, and humane guidelines to help you overcome such common puppy problems as jumping up, mouthing, chewing, house-training difficulties, barking, and submissive urination. KPT may also use creative obstacle courses to help pups build confidence in a variety of situations, like climbing over different surfaces, going up and down stairs, and going through tunnels. In KPT the emphasis

---

*For a fuller discussion, see The Monks of New Skete, *The Art of Raising a Puppy* (Boston: Little, Brown, 1991), pp. 21–60. Subsequent references are to more detailed discussions found in that book.

**Ibid., pp. 61–70, 262–65.

is on fun, but the pups are receiving important socialization at the same time. If you can't find a class, follow the instructions here until such time as you can enroll your dog in advanced training instruction.

## Socialize, Socialize, Socialize

Before you jump into training your puppy, it is a good idea to understand what happens to puppies physically and psychologically as they grow. As indicated, puppies go through critical developmental stages. For the first seven weeks the puppy has special needs as he passes from the neonatal period through the transitional stage and into the socialization period (five to twelve weeks of age). Between one and five weeks he is focused primarily on his mother and littermates. Mother teaches her pups basic dog manners, and the pups learn to interact with one another appropriately. This contact is absolutely essential. It is vital that pups not be separated from their mother and littermates prematurely (say, at five weeks of age). If they do not have this chance to socialize properly with other dogs, such pups are more likely to develop problems involving dog aggression.

As the pup moves into the middle of the socialization period (six to seven weeks), he begins to open up more to people. Contact with human beings is crucial now and needs to continue after you've brought your pup home from the breeder. It is the foundation for your pup's becoming a friendly adult dog. Therefore, if you are buying a puppy, plan on adopting him between seven and ten weeks old, but not earlier.* If the adoption is after ten weeks, make sure the puppy already has been thoroughly socialized at the kennel.

You should also be aware that between approximately eight and ten weeks, a puppy passes through what specialists call "the fear period."** During this time, the puppy is especially vulnerable to stress, neglect, and poor handling, which can leave a lasting imprint. Each puppy experiences this phase differently. It is important to be aware of the fear period, but don't make the puppy live in a cocoon during this or any other period of his early life. Be sensitive to your pup, but don't pamper or coddle him — and don't shelter him. Social

*Ibid., p. 70.

**Ibid., pp. 71, 72.

isolation destroys a puppy faster than any of the most inhumane training methods. Never pet and praise him when he's showing fear at a new experience. Doing so only reinforces the fear reaction. Simply be patient and encouraging. Always praise displays of curiosity and self-confidence.

Socialization embraces the full gamut of a pup's interaction with his world. It goes beyond ordinary encounters with adults and children to include exposure to different environments, surfaces, sights, sounds, dogs — in short, whatever the dog needs to adjust to in his life. The more you introduce your pup to these sorts of experiences now, the easier the pup will be able to adapt later on. Without doubt, if trainers were to single out one behavioral issue owners should concentrate on during the first months of caring for a puppy, it would be socialization. This needs to be emphasized. Don't take any shortcuts in this regard. For the properly socialized pup, the sky is the limit as to what sort of a relationship with you is possible.

At the same time, you can start with a gentle introduction to training procedures. Teaching the puppy to follow, walk on a leash, sit, lie down, and chew only on acceptable objects is perfectly fine during this period, but make sure only capable people work with the pup. For instance, if the pup is brought home during this time, don't allow children who are rough with the dog to handle him until he is ten weeks old, and then only while being monitored by an adult. Some children simply do not have the knack of puppy socializing and training and must wait until the puppy is more mature.

Another facet of socialization involves grooming. Begin grooming sessions with your puppy right away. At this young age there probably won't be much hair to brush, but it is still important to get your pup used to frequent grooming and handling. Since this will become a frequent ritual throughout his life, it is best to condition your pup to enjoy these sessions. Hold your pup by his collar, pass the grooming rake lightly over him several times, and praise him warmly. Next, gently handle and massage his paws and ears, and hold his muzzle briefly as you look into his eyes. If the puppy struggles, lightly shake his muzzle briefly with a curt "nah," but keep the tone gentle. Don't let the pup "win" (i.e., get his way), which would only teach him that by struggling he can get its way. For more detailed guidelines, consult chapter 12 on grooming.

One cardinal rule of puppy training: never lose your temper. These exercises are meant to be fun for both you and your puppy.

## Naming the Pup

In general, we try to name our pups with short, two-syllable names. Names that end on soft *a* or hard *o* are excellent (Sarah, Bosco, Sandra, Laika, Elko, and so forth). Though there is nothing wrong with the traditional Spot, we have had quicker responses from puppies with two-syllable names. Do not use names that rhyme with or sound like obedience commands. Remember that "cute" names may sound quaint on puppies (Cupcake, Huggy, Baby, and the like) but lose their charm when the dog is older. Joke names or names that emphasize a physical characteristic of the breed or individual dog are a matter of personal taste, but generally we recommend against them. Dogs are remarkably sensitive, and often seem to intuit when they are the butt of sarcasm. We also know some city dwellers who have deliberately named their dogs with macho-sounding names in order to increase their protection value (Chopper, Rip, Lance, and Wolf). Again, you have to live with your dog's name.

Once you know your dog, you can even change it. We knew two German shepherd owners who owned dogs with the innocent names of Dagmar and Cain — not exactly terrifying call names. The dogs themselves were mild mannered and easygoing. But when suspicious-looking persons were seen around the home, the owners would call the dogs by shouting, "Killer and Fang, come here quickly!" It proved to be very effective.

Your main concern should be to choose a name that the dog can hear and understand easily, one that complements the dog's breed and personality.

## Coming When Called

First, the pup must learn her name. Say it often, especially when you catch the pup looking at you. A good exercise to help imprint the name quickly is to practice calling the pup back and forth between handlers. At first, leave only about five or six feet between the two handlers. Lengthen this space as the pup becomes more consistent.

*Call the puppy between two handlers. Open your arms wide to receive the pup as you squat or kneel down.*

*You can use a long leash to guide the puppy between two or three people. After the puppy comes to you, throw the leash to the next person before he calls the puppy in.*

Call the puppy in a light, happy tone of voice, and when the puppy comes to you, praise her exuberantly. You can also use a treat to enhance the motivation. You should be on your knees when you call the pup. Your arms should be open wide, to help "funnel" the pup toward you. If your pup does not wag her tail and have a happy look in her eye when coming to you, perhaps you're not using enough animation and lilt to your voice when you call her name. Don't take it personally if a pup is slow coming to you. Continue calling the puppy and give her encouragement the minute she starts to move toward you.

Keep calling if the pup doesn't come. She may be confused. Pat the ground hard, click your fingers, clap your hands, or jingle a set of keys. When the pup comes to you, praise her physically and verbally. Face the puppy toward the other handler to prepare her to return. A helpful strategy is to gently restrain the pup as the other handler calls her enthusiastically. The pup will start to protest and try valiantly to respond to the recall. Let go of the pup, and when she runs to the other handler, have him give the pup a treat and praise her enthusiastically.

If your pup is coming pertly between two handlers, you can go to circle-come work. Add another handler, and space yourselves about five feet apart. Attach a light lead to the pup's collar, have one of the partners call the pup, then toss the end of the leash to him. If the pup does not come quickly, have that partner give the leash a light tug to initiate the recall. Usually one tug is sufficient; however, if the pup stops, tug again as necessary. Don't pull the pup. You want the pup to make the decision to come. Have the pup sit in front of you when she comes. Some pups may jump up and paw each handler. Do not discipline this behavior, but gently ease the puppy to a sitting position and continue praising her. Remember, no punishment should ever be connected with the action of coming. Continue having the puppy called around the circle. Puppy call-in sessions should each last five or ten minutes until the pup is five months old. After that, sessions can be extended. End your call-in sessions with some play. Leave the leash on during the play period. Doing so helps the pup to feel comfortable with the leash on and to associate wearing it with pleasant experiences, not just "work."

When you call your pup, use the dog's name and the word *come*. Do not use more than one name, and do not use affectionate nicknames. The more you work on a prompt recall in your dog's early days, the fewer recall problems you will have later on. Almost every dog has on occasion failed to come when called, but puppies who have experienced coming when called as a happy experience early in life tend to have a better attitude toward the recall later.

## Following

Teaching your pup to look to you as a humane leader is important from the very start. During the first couple of weeks the new puppy owner should have an off-lead follow session with the pup once a day, for ten to twenty minutes.* Take the pup alone to an area where there are few distractions. Set the pup down and slowly walk away, keeping the pup's attention by talking in an encouraging, animated tone. Say the pup's name frequently. Stop every so often, crouch down, and praise the pup. Then rise and begin walking again. Make plenty of turns in your follow session. Trace a figure eight every so often. Most puppies, from the fourth week on, naturally follow a moving human being in much the same way as they would tag along with their mother. Often no coaxing is necessary, and the puppy may even be underfoot. This is fine — the point is to keep the pup with you, as near as possible. You may want to use a set of keys, hand clap, squeak toy, or whistle to help orient the pup to you. Don't use food for this exercise. Keep moving away in one direction steadily and you'll find the pup trotting after you. This helps get the pup's attention in a natural way.

Although this session might become boring in time and seem uneventful, keep it up for at least three weeks after you get your new pup. Doing so helps build rapport in several ways. It helps the pup recognize you as leader, the one to follow. The pup's being comfortable and happy when following you will help eliminate recall problems in the coming months, when your puppy will enter a more independent stage and be more easily attracted away from you.

## Leash Work

Leash work can begin as early as six or seven weeks of age. The revealing term *leash-breaking*, often used by trainers to explain how a puppy should be introduced to the leash, is unfortunate, since there should be no "breaking" of the pup involved. If you practice the following exercises, leash work follows more easily. First, accustom your pup to wearing a collar. Use a flat leather buckle collar if possible. As you hold the pup in your arms, put the collar on snug once or twice a day, to get the pup used to the feel of the collar on her neck.

Start familiarizing your pup to walking on a leash by attaching a light leash to her collar and letting her drag it along on a normal

*William E. Campbell, *Behavior Problems in Dogs* (Grants Pass, Ore.: BehavioRx Systems, 1999), p. 172.

*Keep holding the leash up high in early leash training.*

*Call the pup toward you with a gentle tug on the leash. Use a pleasant voice so that the pup comes and gets used to a leash.*

follow session. Don't use heavy metal leashes on puppies. After the dragging phase, the trick is to attach the leash and begin walking without the pup's knowing it is on. As you pet the pup, snap on your lead. Begin to walk, and hold the lead almost straight up, out of the pup's way, and don't apply any pressure to the lead. At some point, the pup will stop and feel the lead. Don't pull the pup along. Stop, bend over, and say, "Okay, Sasha, let's go!" Use the same animated tone of voice you have used in your follow sessions. If the pup balks, stop for a minute and reassure her. Some pups brace themselves dramatically and become quite vocal. They are usually pups who have had trouble sticking close on the follow exercises. You need to tug the lead gently and encourage the pup until she decides to come along. Some researchers have found that males tend to be more feisty on initial leash work, running ahead and vocalizing when they hit the end of the leash. Whatever the reaction, get the pup focused on you and begin again. Make your first sessions short, and try to end on a happy note. Leave the leash on and have a play session with the puppy.

Now move to the next level, using a retractable leash. This is an excellent device to get your pup used to the feel of the leash without any compulsion. The pup learns to be comfortable walking on leash and to handle an occasional light tug without feeling the constant pressure of the leash on her neck. As soon as the pup shows any resistance, simply release the pressure on the leash and pat the side of your leg encouragingly as you call your pup. Before you know it, your pup will be walking with you in a relaxed manner.

As soon as the pup is comfortable walking with you, tie the leash to your belt loop and have the pup follow you around the house naturally throughout the day. This is an excellent bonding formula and reinforces the initial adjustment to the leash.

## Teaching the Sit

Puppies should be taught to sit using a nonphysical technique, employing treats or some other object of attraction. First, get your pup's attention by showing him a small treat, holding it a little above eye level just in front of him. As he looks at your hand, raise it slightly over his head as you say, "Sit." Most likely the pup's rear will go down

*Use the pup's natural energy and curiosity to guide him to sit on command.*

simultaneously. Praise the dog gently until he is fully seated, and then give him the treat. If the pup does not sit the first time, try again. If the puppy is simply looking up and not sitting, repeat the procedure, and with your other hand, lean over and lightly touch the pup's rump. As the pup sits, give him the treat. We have found this method of teaching the sit to be faster and gentler than leash and training-collar methods. After the puppy sits, keep him in place as you praise him.

Practice having your dog sit five times in a row per session, and have a session at least once a day for a week. Also, take advantage of daily feeding times to have your pup sit for his food using the same technique. As you see that your pup has mastered the command, start weaning him away from the treat by giving the treat every other time, while continuing to praise him. This quickly allows you to give a treat to your pup randomly. He doesn't know when and if the treat will come, an important incentive to keep listening to you.

## Teaching the Down

Like the puppy sit, the puppy down is best taught with a minimum of force and compulsion. Since the pup already knows how to lie down, what you are teaching him is *when* to lie down. To aid in teaching this exercise, use a treat, ball, or other some other toy the pup likes. Start with the pup in a sit. It is easier for a dog to lie down from a sitting position than from a standing one. Show the pup the treat, and as he focuses on it, lower it directly to the ground about six inches in front of his feet. As you do so, give the command "down." As the pup lowers his head for the treat, praise the pup gently, repeating, "Down," if necessary.

If your pup is a little older (sixteen weeks), crouch down with him to your left as he sits. Put several fingers of your left hand under his collar along the side of his neck. With a treat in your right hand, show it to him by moving your hand directly in front of him and then lowering it straight down to the ground, applying very gentle downward pressure on the collar as you do so. Your pup will follow the treat down. When he is in the down position, praise him and give him the treat. Keep your left hand in the collar as you continue to praise him so that the pup learns not to get up immediately. As with sit, practice the down in sets of five at least once a day for a week or two and gradually wean him off the treats, as you did with the sit.

Following these guidelines allows the pup to learn the down in a low-key, nonthreatening way. Most puppies respond well to this method. As the pup shows understanding, you can begin to extend the length of the down by practicing in everyday situations. Always keep it positive and upbeat. Bear in mind that the physical methods described for older dogs in this book are not recommended for puppies, because they involve firmer body contact and corrections. Never try to train a puppy to lie down using the "pulley" method of stepping on the leash and forcing the dog down.

As with teaching the sit, limit puppy down sessions to five or six repetitions. One session in the morning and one in the evening is sensible.

## Tethering the Pup at Night

The best place for a puppy to sleep is the floor at the foot of your bed. Avoid socially isolating the puppy at night. For the first few nights, an

*Lower your hand to the ground to teach the puppy the down.*

*The pup will follow your hand down; as he does so, say, "Down," and praise him.*

old sheet or blanket folded up next to your bed works well. You can also use a cardboard box with low sides to create a comfortable, den-like resting place. If a breeder has already accustomed a pup to a shipping or open-wire crate, you may wish to use it; other owners prefer to tether their pups for the night. Tethering can help prevent overnight accidents (since most puppies do not soil their immediate area) and curb destructive chewing. The tether can be quite short at first — a foot or two will do. Use a metal tether. The puppy could chew on any other material and eventually free himself, possibly in the middle of the night. The pup will probably fight the tether the first night. Reassure him without coddling him. You might try attaching the pup before you actually turn out the lights, to let the pup get used to the tether while you read. Above all, don't give in and release him. Some initial whining and protest is inevitable. The pup will settle down eventually.

Attach the end of the tether to a strong, immovable object, such as the leg of the bed itself. Make sure that the tether cannot break and that it is not attached to an object that can be pulled away by the pup. Puppies are strong little critters when they want to be. One man attached his puppy to a hamper on wheels and retired for the night. About three in the morning the puppy pulled the hamper to the top of the stairs and sent the hamper and himself crashing to the bottom. We heard about the incident when the owner later brought the puppy to us to learn how to climb stairs, a skill he had not yet acquired. A little research pinpointed why the dog had such an aversion to stairs.

In general, our practice is to keep our pups tethered or confined at night until two objectives are accomplished. First, the pup must be fully housebroken. Second, all chewing, digging, and other puppy mischief must have ceased. Only then do we allow the pup freedom in our bedrooms at night.

## Behavior at Night

Since our monastic bedroom section is in the cloister, where silence is required, no puppy yodeling or barking is allowed. Since dogs naturally bark at suspicious noises or the sound of an intruder, you don't have to encourage alarm barking at night. If the pup starts to carry on, simply shush her after a few seconds. Chapter 20, "Where Is Your Dog This

Evening?" describes "bedroom etiquette" for adult canines, which applies to puppies as well, but there are some exceptions. If your pup pipes up during the night, she could be telling you that she needs to go out. Young puppies who are not fully housebroken need a trip out approximately every four hours. As the pup matures, after twelve weeks of age, she should be able to pass a night of eight hours' duration without needing to go out, assuming she has not been fed or given water near bedtime. Many a pup owner gets into the "three A.M. trip outdoors" ritual needlessly. The pup whines, the owner gets up and takes her out. But if you have recently let the pup out and are sure she is "empty" or relatively so, shush her with a sharp "No, go back to sleep." Don't become a doorman for the pup. At some point, the pup must learn control. If there is an accident, don't panic (see below, "Asking to Go Out").

When you tether or crate your pup for the night, provide her with some toys and chewables. Remove food and water. If you are housebreaking the pup, remove water at least four hours prior to retiring and feed the puppy her third meal late in the afternoon. This early feeding gives the pup time to eliminate before settling down for the night. Additionally, most pups can move from three to two meals per day between sixteen and twenty weeks of age, which also helps stop the need for a late-night potty break.

## Using a Dog Crate

A dog crate can help you train a puppy in a variety of ways. Dogs are basically den animals, and a crate approximates the den of their wolf ancestors. Crates are not cruel or inhumane, as long as they have ample room and are not used as punishment. We have found that dogs enjoy them. To use an analogy, many parents consider a child's playpen invaluable. We prefer airline shipping crates made of molded plastic with metal doors. They are safe and comfortable, and provide a secure denlike atmosphere for your pup while letting him see out of all sides. They are also easily cleaned if your pup ever vomits or has an accident in it. Open-air metal crates, though collapsible and convenient, have sharp edges that a pup could get a collar caught on. They also are not as denlike as the shipping crate. If you use an open-air crate and your pup doesn't seem relaxed, cover the three sides and top with a blanket.

*Dogs and pups easily adapt to their own den/crate at night or for a few hours during the day.*

It is a good idea to accustom your pup to the crate early on. Begin by allowing him to investigate the crate without closing the door. Give him a treat and praise him warmly if he goes in. If the treat does not induce him to investigate, put him in and give him the treat. Do this several times till he feels relaxed about going into the crate. Next, have him go into the crate and close the door for a brief period as you remain close by. Offer him a treat. Give him a favorite toy. Work up to the point at which the pup can stay a couple of minutes without protest. Use a treat as a reward. Finally, make the transition to feeding the pup in his crate, which deepens the positive association with the crate and lets you gradually increase the length of time the pup can stay in the crate on a day-to-day basis. Just remember that the pup will need to go out to relieve himself soon after eating. Many owners take the precaution of removing their dog's collar during the time in the crate, to prevent any possibility of it getting caught.

At times some pups can be stubborn about the crate. If your puppy barks in the crate (and you are sure he doesn't have to go out), go to the crate, clap your hands, and tell the pup, "No, no barking." If the pup continues, use a squirt of water from a spray bottle or a shaker can filled with pennies, which many pups dislike the sound of. If all else fails, rapping the top of the crate sharply with your hand, coupled with a "no bark!" has often been effective.

You can use the crate at night or when you will be gone for a few hours. Since most dogs are fastidious about their immediate area, you can use the crate to help the pup spend time alone without relieving himself. The crate is also a temporary solution until he is house-broken and can be trusted alone in the house while you are busy, sleeping, or gone.

Later, you will find further uses for the crate. Many dogs prefer to ride in a crate in the car. When you are moving or carrying a large amount of baggage, a crate can save valuable space. It is the safest place for a dog in a car accident. If you plan to travel with your dog, it is wise to crate-condition the dog in advance. When we raise litters, we let the pups use an open crate as a playhouse, so that they are ready to use a crate in their new homes. Never ship a dog in a shipping crate on a train or plane unless you have first conditioned the dog to the crate. If you do need to ship, use the smallest crate possible. If the crate is too large, the dog gets bounced around in transit. With a small crate, the animal can brace himself against the walls.

## Asking to Go Out

Our procedures for house-training puppies can be found in chapter 39 on house training and in chapter 11 of *The Art of Raising a Puppy.* Asking to go out is best taught in puppyhood and is a service any canine owner appreciates. If you want a puppy who asks to go out, begin early by recognizing the signs the puppy gives indicating she wants to go out to eliminate. First, anticipate the pup needing to go out between five and fifteen minutes after eating. Some pups clarify this by barking or whining. Be sure to respond by opening the door.

With a pup who does not vocally indicate a desire to go out, look for other signs that elimination is forthcoming. Some pups pace, noses to the ground. Others may go to the door and scratch it. Still others may come to you and jump up. Whatever the sign, read it quickly and take the pup to the door, asking in an excited tone of voice, "Do you want to go out?" Make sure you use the same phrase and the same door each time you let the pup out. If the pup lives in a family, everyone should be consistent about what phrase is used when inviting the pup out for elimination. But don't let the pup out each time she nears a door. Pups do not need to go out every hour on the hour, and you should be careful to avoid the extreme of becoming a doorman for the pup. Though asking to go out is certainly a convenience, don't let the pup overdo it.

Pups become conditioned to regular routines quickly. If you feed her at the same time every day and let her out on a frequent schedule, her elimination needs become fairly predictable. For an older pup, when you are in the middle of preparing a meal or in some other situation when you cannot immediately let the dog out, say, "No, wait!" and let the pup out as soon as you can. At night, follow the instructions for silencing whiners mentioned earlier. If your pup does have an accident, follow the procedures outlined in the house-training section. Clean up the area with disinfectant but apply a final solution of white vinegar and water, which cancels out the odor from the dog's point of view.

## Submissive Urination*

Submissive urination is different from house soiling because it is involuntary and usually takes place when the puppy is in a subordinate position and releases a puddle, or possibly a flood, of urine. The puppy does not mean to urinate but rather to show submission. This behavior has its roots early in puppyhood, when pups were initially stimulated to urinate and defecate by their mothers. Gradually this behavior came to be expressed as a sign of submission to the mother or other older dog or person. Though the behavior invariably stops with time, here are some techniques to help you through the problem.

Above all, don't punish the puppy. Try not to tower over the pup in a threatening manner. In general, crouch down to the side when you call the pup, and praise him by stroking him under the chin and on the shoulder or chest region. Experience will teach you on what occasions your puppy tends to urinate involuntarily. Greetings and departures, visits of company and relatives, or wild activity by children often trigger submissive urination. Don't isolate the pup from

*See *The Art of Raising a Puppy*, p. 212.

these situations, but first try to delay the submissive response by simply ignoring the pup and avoiding eye contact. As the excitement of your arrival dies down, let the pup approach you and offer him the palm of your hand, still avoiding eye contact. Gently pet him under the chin and go about your business. Keep your greetings low-key. Keep active children away from the pup, but invite gentler children to play with him. Puppies who wet submissively for adults often do not for children, so it may be helpful to find a placid child willing to take the pup for walks and socialize him.

Gradually build the pup's confidence by taking him for walks on busy sidewalks and in crowded areas. Avoid any sort of correction for submissive urination.

## Discipline for Puppies*

Puppies rarely need strong physical corrections. Some situations merit discipline, and we suggest that you use the puppy shakedown method, which is described in chapter 10. Kindergarten puppy training is designed to eliminate the necessity of giving serious physical corrections later on in life. But every dog occasionally gets out of line, and you should not hesitate to use an appropriate physical correction for big offenses, such as biting and nipping or house soiling (assuming you catch the older pup in the act and she has been conditioned properly to house-training etiquette). Different breeds need different approaches. German shepherds, Dobermans, and rottweilers can often get themselves into situations in which physical discipline is necessary. Terriers can be highly kinetic and often need sustained eye contact and only occasional physical discipline. You must "read" your own dog.

## Puppy Chewing**

Almost every puppy, at some point, chews something she is not supposed to. It might be an old shoe or it might be a precious family heirloom, if you are naive enough to leave one within reach. Before you bring your puppy home for the first time, "chew-proof" your home as much as possible. Make careful arrangements for shoes and socks to

*Ibid., pp. 201–05.
**Ibid., pp. 206–08.

be up off the floor, throw rugs stapled down or removed, and valuable objects put away, at least until the puppy is past her chewing stage. Use a crate when you are not home or are unable to monitor your pup. At other times, provide your pup with a blanket or dog bed where she can go when tired, preferably in two locations, one in your living room (or den or kitchen) and the other in the bedroom. In these areas, provide the pup with a rawhide or nylon bone. Meat-scented and rope bones are excellent, too. Do not use real bones. Stay away from cute toys sold at pet stores, or contraptions that can be broken or chewed apart. Make this bone the center of attention for the pup by playing fetch with it, wiggling it on the floor and letting the pup chase it, and giving it to the pup whenever she begins to nibble on a forbidden object. This is the Toy. Get a few different sizes of play bones so that you can graduate to the size as the pup grows older. Before leaving the pup — whether confined to a crate or a small area such as a kitchen, or left outside — roll this bone between your palms for a couple of minutes to leave your scent on it, and dramatically present it to the pup as you leave home. Another method of orienting the pup to the bone is to rub your saliva on it. If you plan to take your pup for a ride in the car and leave her there, take the special bone along, too, and present it to the pup as you leave the car.

Whenever you catch the pup chewing on something forbidden, say a deep and strong "nah!" and immediately take the object out of the pup's mouth or remove the pup from the scene. Immediately get the play bone and present it to her, saying, "Here, this is for you, this is your toy." Avoid punishing the pup or trying to tear forbidden chew-ables out of her mouth. Avoid playing tug-of-war with the pup to remove an object from her mouth, which can lead to possessiveness later on. To get an object out of a pup's mouth, open the jaws by plac-ing your hand across the top of the muzzle behind the teeth, thumb on one side and fingers on the other, and with the other hand pull down the bottom jaw. As the pup drops the object, say, "Good!"

One common response to a chewing problem is to provide the pup with myriad chew toys, in the hope that the pup will find this vast array satisfying and stay clear of taboo objects. Though somewhat log-ical from a human point of view, this technique usually backfires because it conditions the pup to perceive practically everything as a potential chew toy. Provide one or two toys and stop there. If you must have several toys for several locations (car, bedroom, basement), make them all the same kind of toy.

The pup needs to chew. Never completely forbid your pup to chew, but rather focus her attention on the proper chew articles. Not to allow chewing is unnatural and ill advised. The pup's proper oral development will be stunted, and since the pup was never allowed to chew on anything while going through the normal teething stage, she may decide to postpone oral activity until later on. The renewed activity could include destructive chewing of objects and possibly humans. As a stage, chewing can last as long as eight months, or as short as four months, depending on the individual dog.

## Praising Your Puppy

Physical and verbal praise are important for puppies. They are the cornerstone of a good relationship and must be deliberately cultivated. Aside from ordinary play with your pup, such as fetch and the like, include a play session with both of you on all fours at least twice a week. This keeps your relationship relaxed and playful. Keep the pup animated and happy during formal obedience work, too.

At the same time, avoid any tendency to coddle the pup. An overindulged pup is soon spoiled and takes advantage of the rest of the family pack by assuming a leadership role. We have seen puppies who are dictators. When they whine, they are picked up. If they scratch a door, they are immediately let out. If they bark at someone, the owner pets them and thanks them. If they bite someone, they may or may not be scolded, depending on what the owner thinks of the victim. If the puppy wants praise, he need only nudge or jump up on the owner to be given attention immediately. The owner who responds to such behavior by coddling and cuddling teaches the puppy that such behavior elicits affection. The puppy learns to turn on this behavior whenever he wants attention, which may be every other minute.

A more sensible approach to praising your puppy includes praising the pup for good behavior and avoiding the temptation to praise the pup simply because he is there. While affectionate petting and squeezing are a part of any initial puppy-owner relationship, decrease this behavior as the puppy matures and begins training. Praise the pup when he fulfills a command, like sit or come. It takes only a second to ask the pup to sit before praising him, and it will be well worth your while, for it orients the pup to obedience early in life. You can do so easily without compromising the overall attitude of playfulness and fun. The last thing puppies need is a martial regimen, but you can still keep praise meaningful.

## Lonely Puppies

Patience and a sincere desire for constant companionship are the two most important traits necessary to raise a healthy, happy puppy. If you are the kind of person who needs long periods of time completely alone or who feels occasionally claustrophobic if even a dog is around, don't get a dog. A dog will always be there, asking for attention and directions, asking to serve. If this gets on your nerves, reconsider whether you want a puppy. There is nothing wrong with admitting that cats are your type of pet or that a bowl of goldfish would adapt better to your lifestyle. Respect for living things requires an appreciation for their innate needs and desires and your own ability to fulfill these, at least to some degree.

A belated realization of the above sometimes leads to puppies with behavioral problems. Puppies can be bored, neglected — in a word, lonely. Puppies simply cannot endure long periods of isolation. They are pack animals, born and raised in a litter. They need social experience throughout puppyhood and later life. Lonely puppies are puppies who are left alone for nine to ten hours while their owners work. They are pups who are refused admission to the bedroom for sleeping and are banished to the kitchen, basement, or garage (often on the pretext that they must "protect" the house — something they will never feel compelled to do unless they are first endeared to the masters of the house). Lonely puppies often lack play experience with humans or other dogs. Instead, their isolation is compensated for by continual stroking and cuddling in the arms of a guilt-ridden owner. Such touching has little effect on the core reality of loneliness. The best food, trinkets, treats, and even professional training does not matter if the puppy lives a hermit's life.

Lonely puppies vent their frustrations by chewing, digging, barking, whining, and scratching doors and walls. They crave human contact and seek to escape their isolation, even though they may appear shy or aggressive around humans at first. If you properly socialize your puppy from the beginning, he should not be lonely, by your standards or his. If, however, you decide to modify your pup's lifestyle after deciding that he can "take it," and isolate the pup for significant periods of time, you will most probably wind up with serious behavioral problems.

What it all adds up to is this: don't get a pup unless you have plenty of time to spend with him. Don't expect a dog to be a cat or a hamster. Answer the following questions honestly if you are planning to get a puppy. If you already have a dog, these questions might serve as a helpful review of your responsibilities as a dog owner.*

Will someone be home to provide meals for the animal according to a fixed schedule?

Will someone be home during the day to look after the dog?

Are you willing to exercise the puppy at least twice a day, according to a set schedule?

Are you willing to secure proper obedience training for the dog?

Are you willing to pay for all inoculations, periodic veterinary exams, and any emergency treatment the dog may need?

Are you willing to pay the cost of spaying or neutering your pet to prevent the birth of more unwanted puppies in a nation already saturated with pets?

Are you willing to obey the license and leash laws of your community?

Are you willing to groom, bathe, and pick up after the dog?

Are you committed to caring for the dog for its lifetime?

*These questions are adapted from "A Checklist for Potential Pet Owners" put out by the Humane Society of the United States before this book was first published. They are as relevant today as they were in 1978.

Getting a puppy is a joyous event, but one filled with real responsibilities. We make no apologies for ending this chapter on a serious note: each year too many pups go into homes in which owners have not thought through the commitment entailed in caring for a pup. Get advice. Think carefully.

# Standard Obedience Exercises

# 30

# Starting Out

The following chapters will help you teach your dog how to heel, sit, sit and stay, stand, lie down, lie down and stay, and come when called. As the exercises are presented, we presume your dog to be at least six months of age, having received at least some exposure to elementary puppy training. In our experience, the most popular exercises are the last two. Most dog owners are happy if Fido simply lies down or comes when he is called. But in order to teach these two exercises effectively, your dog must know something about the others. How reasonable would it be to expect a child to tackle algebra before she's mastered addition and multiplication, division and subtraction? The former is built on the latter. As we show in the following chapters, heel, sit, stay, and stand are not strictly ornamental — they can be used to great advantage and are an essential part of your dog's foundation in training. We will try to explain fully what each exercise entails and demonstrate at least two methods of teaching it. We realize that each dog is an individual and that not every dog responds well to every method. Finally, we include examples in each chapter of how the exercise can be used practically in your day-to-day life with your pet.

We emphasize the practical application of each exercise, and not "showing off" your dog to strangers with clever parlor tricks like shaking hands, "praying," rolling over, and the like. These tricks may be entertaining, but the dog who can shake hands, roll over, and pray but cannot lie down and who runs away when called can be a problem to his owner.

Nor is our emphasis on training for the professional obedience ring or for competition. Although what we say does not contradict training methods for the ring, this book is geared toward companionship, to help you read, teach, and manage your dog in everyday life, thereby deepening the relationship both from your perspective and from your dog's. If you are interested in showing your dog in the obedience ring, we suggest that you consult some of the many books that concentrate on that field (see Select Reading List). Nevertheless, even those who show their dogs professionally need to stop occasionally and reevaluate their relationship with their dogs. Regardless of how well a dog knows his exercises, the total relationship between dog and master remains the most important aspect of having a dog as a companion.

Here's a story that illustrates what we mean. A famous dog, well known for his high scores in the obedience ring, arrived at the show grounds. As we watched, the owner carefully opened the dog's metal crate, warning the dog not to barge out. Instead, the dog nudged open the door of the crate, practically knocking over his owner as he barreled out of the station wagon. Luckily, the owner had left the dog's leash on and pinned it with his foot, thereby preventing the dog's escape. He immediately snapped the dog to a heel position and began drilling the dog in the obedience exercises. The dog responded like a robot. He racked up a high score in the ring. Immediately after the long down, which is taken in a group with other dogs, the owner leashed his dog and heeled him out of the ring. But once outside the ring rope, the dog lunged ahead, practically dragging his trainer across an empty field, back to the station wagon. The dog leaped into the crate in the car, and the trainer, exhausted after the "ride," slammed the crate door shut. Though the dog performed well, one wonders what the total dog-human relationship was like.

This story is not typical of most dogs shown in the obedience ring, but it does point out a major pitfall in obedience-training exercises: don't make your dog into a robot! Obedience training does not turn a dog into a zombie, but a bad trainer does. Train with spirit, humor, a judicious use of treats, and, most of all, physical and verbal praise.

# 31

# Equipment

What kind of equipment do you need to train your dog? First, you will need a quality leather or cotton webbed six-foot leash. Here at the monastery our preference is for leather leashes because they are durable and, once they are broken in, extremely comfortable to work with. However, there are trainers who prefer cotton webbed leashes, believing that they have less give, thus making them capable of eliciting a quicker response from the dog. Avoid nylon leashes, which remain stiff and can burn your hands if your dog lurches unexpectedly, as well as metal leashes, which are entirely unsuitable for training. However, you may wish to have on hand a twenty-six-foot "flexi-lead," or

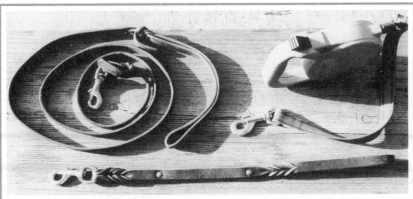

*A good lead has no sewn parts that can break. Braided leather or webbed construction works well, with a heavy-duty clip. Shown is a short lead, a six-foot-leash, and a retractable leash (helpful outside of training sessions for informal walks).*

*We prefer a braided nylon training collar. The Volhard collar has a floating ring and clip — particularly useful for a precise fit. The prong collar looks worse than it feels and is helpful for large, unruly dogs.*

retractable leash. Although you won't use this for formal training, such devices are handy for informal walks with your dog.

The width of the training leash depends on the size of the dog you are working with. Extremely large, oversize dogs can use a three-quarter-inch-wide leash. For the average to larger-size dog, we recommend a leash that is half an inch wide. For smaller dogs such as the toy breeds and some of the terriers, use a leash that is a quarter inch wide.

In addition to a flat-buckle collar for dog tags, you need an appropriate training collar, of which there are several possibilities. Most people are familiar with steel or nylon slip-on collars, which are made from a piece of metal chain or flexible, braided nylon attached to two rings. Readily available at most pet stores or through pet-supply catalogs, they slide easily over a dog's head but often have the disadvantage of being too large for the precise size of the dog's neck. Ultimately, this can minimize the effectiveness of a leash pop. In addition, recent research has shown that metal training collars can also cause damage to the dog's trachea if used incorrectly. However, if you do use a steel collar, make sure the links are small, pounded flat and not rounded, so that the collar has good, clean action when pulled and does not get hung up, hindering a smooth release.

A better option for most dogs of normal touch sensitivity is a nylon snap-around collar, which has been popularized by Jack and Wendy Volhard. A snap-around collar consists of a piece of nylon with a clasp on one end and a ring on the other, with a loose, or floating, ring sliding between them. Though not as easy to find as slip collars, they have the advantage of being able to fit precisely to the exact size of the dog's neck and can be used as a safe flat collar when the dog is not being trained. We list where they may be obtained in the back of this book. Keep in mind, however, that certain longhaired breeds do not do well with any sort of nylon collar (the hair tangles in them) and need a metal slip-on collar to allow the collar to pass through the coat smoothly.

A third possibility is to use a prong (sometimes called a pinch) collar. These collars look somewhat medieval and can inspire skepticism in a sensitive owner who doesn't wish to harm her dog, but we have found them to be safe and humane when used correctly, particularly for touch-insensitive dogs. The individual prongs come apart so that a

*When using the special Volhard collar, attach the clip from under the neck to the floating ring, then connect the leash to the end ring.*

precise fit can be achieved — snug — right under the ears. In fact, many dog professionals and even chiropractors believe them to be the safest training collar, and there is no question of their effectiveness. The principal drawbacks to using a prong collar are that it is more difficult to put on than a slip collar and that it cannot be left on when not in use. Prong collars come in three different link sizes: small, medium, and large. We use small and medium collars for all dogs; the large collars are bulky and less efficient.

A final point about two other collar options: recently head halter (Halti) collars have become popular with some trainers. These fit over the dog's head and muzzle and are intended to guide the dog by controlling the movement of his head. They work on the assumption that the dog's body follows the position of the head; however, we have found that most dogs object to them vigorously, especially in the early stages of training. Further, they control the dog without really teaching the dog to assume responsibility for his behavior and can potentially damage a dog's neck if the dog or handler pulls suddenly. Finally, we do not find electronic collars suitable for basic obedience training. We prefer them for more advanced training that is done off leash; they require a good bit of skill, experience, and commitment to use effectively.

Whatever training collar you use, make sure it fits the dog properly! The most common error clients make in "outfitting" their dogs for training is to purchase a training collar that is many times too large. Not only is this dangerous, the more oversize your training collar, the longer it takes for your correctional tug to be telegraphed to your dog when you use the training leash for a correction. You owe it to the dog to make your corrections instantaneous, and you can't do so with oversize equipment. *Measure* your dog's neck beforehand. Slip-on collars are usually sold in even-inch models, so select one that allows three or four inches of slack when it is pulled tight on the dog. This means that the training collar will be snug going over the dog's ears when you put it on. Both prong and snap-around collars should fit precisely the size of your dog's neck directly under the ears; with the snap-around collar, you should be able to slip two fingers between the neck and collar on a larger dog (one finger for smaller ones) when he is relaxed.

There is definitely a right way to put each of these training collars on. For a slip collar, stand to your dog's right, facing forward. Hold

*The correct way to put on an ordinary nylon training collar. Notice that the collar falls into the letter* p.

the collar out in front of your dog so that the looped portion falls naturally into the letter *p*. Slide the collar over the dog's head, making sure that when you pull the collar, it loosens up when you release the ring. If you have the collar upside down, it will not release. For the snap-around collar, begin by facing your dog with the clip in your left hand and the remaining two rings in your right hand. Place the collar underneath the neck and bring the ends up to the top of the neck. Attach the clip to the movable ring.

We feel strongly that a dog should wear a training collar whenever safety considerations do not prevent it. Owners should be aware that whenever a dog is wearing a training collar, there is the remote possibility that it could get caught on something (for example, a branch or a sharp piece of metal from a cyclone fence) and accidentally choke the dog to death. Hence, we would not recommend leaving the collar on while you are at work or when the dog cannot be monitored. At the same time, we've found that the advantages of wearing the collar when you are around are significant. Once the training collar is on, it can serve as a "mini-leash" when the handler simply inserts a finger in the active ring.

*You can adapt a reinforced cotton rope for recall work. Securely attach a heavy-duty clip to one end. When training with the long rope, wear gloves to protect your hands from rope burn.*

You also need a long cotton webbed leash (twenty feet) for teaching controlled walking and a long rope (thirty to fifty feet) for work in teaching the recall. You can easily make one yourself by taking a length of clothesline and tying a clip to the end of it, which can then be attached to your six-foot training lead. If your dog has a tendency to take off when you call her, wear gloves to prevent your hands from being burned when the rope is pulled through fast. Stay away from twine or string. It can break or cut your hands if the dog pulls away quickly.

Lastly, if you are using treats, we suggest either small cut-up pieces of hot dog that you prepare yourself or, if you are looking for a commercial product, freeze-dried liver treats. Dogs love them, and in addition to your not having to worry about them spoiling, they are not greasy or messy. It is helpful to have something akin to a carpenter's apron to hold the treats while you are working with your dog.

If you want to find quality dog-training equipment, call an obedience instructor or an obedience school, or consult the references for dog equipment at the end of this book. Don't rely on items purchased in supermarkets to last. As the pet market continues to boom, there is plenty of poor-quality pet equipment available. Remember, a training collar that breaks in the middle of a training session can cost you valuable training time. A defective leash that snaps when your dog bolts at a cat on a busy city street can possibly cost the dog his life. Get good, top-quality equipment.

# 32

# Heeling

The object of the heel and the automatic sit is to train your dog to walk on your left side without pulling ahead or lagging behind — and to glide into a sit when you stop. We teach this exercise in stages, first making sure your dog (let's call her Una) is able to walk on leash, then teaching her not to pull, and then, finally, moving into a formal heel accompanied by an automatic sit when you come to a stop. Heeling is a practical, useful skill. It is not an ornamental part of dog training, reserved for professional trainers. When Una walks at your side (and at your pace), she implicitly recognizes you as alpha. Since walking your dog is an essential aspect of your life with her from the day you adopt her, heel (broadly understood here as controlled walking *and* a formal heel) is the first lesson we cover. In many ways how you teach this exercise may be instrumental in changing the way you and your dog relate to each other, particularly if you have not covered the preliminary leash work described in the section on puppy training. This exercise is vital in reordering faulty dog-owner relationships in which the animal perceives himself as the leader and, as a reflection of that perception, takes the owner for a walk every day. If your dog fits this description, and if your arms are continually being popped out of their sockets by a lunging beast, you will appreciate the necessity of this exercise.

Trainers disagree on how the heel should be taught. We try to view this exercise incrementally as part of the overall relationship between dog and owner, and not in punitive terms. Leadership is the issue, so a certain amount of firmness is in order. For preliminary walking on leash and teaching an untrained dog not to pull, we begin with a

*Begin leash training with an older dog by walking in a square pattern with a twenty-foot leash.*

*Briskly set out walking on the first leg of the square, giving the leash a quick pop if needed and not saying anything.*

*Keep moving in a square pattern, and the dog has to follow. Pause a few moments at each corner before you turn.*

*Finally the dog gets the message and tries to catch up with you.*

"square pattern" in which there is no eye contact and limited praise, with the handler moving steadily along the outline of a square (each side about sixty feet in length), pausing for a couple of minutes at each corner. We discuss this procedure in detail below. Once the dog is ready, we move on to the next levels of controlled walking and formal heel, in which we give plenty of encouragement and use treats judiciously. If we progress patiently and methodically, much of the harshness characteristic of certain approaches to heel can be avoided, making for a much smoother learning process. In this spirit, a word of advice: always remember to keep your attitude positive. Determine from the beginning that you are going to learn to work as a team with your dog and that you will eventually be able to walk smoothly together. Don't go into the exercise saying, "My dog can't heel." In our experience, we have never met a dog who cannot heel.

When first starting out, select an area that has good footing and is large enough to permit walking in any direction. Wear shoes that are comfortable and have good support and traction. (Don't try to train your dog to heel when you are wearing sandals and walking on grass!) The area should be quiet — although distractions are important in later training, preliminary sessions are no time for them; make sure there will be no interruptions.

To teach Una to pay attention to you and not to pull on the leash, begin by using a twenty-foot cotton webbed leash, clipping it to the live ring of a correctly fitted training collar. Put your right thumb through the loop and clench your right hand over the leash. Place your left hand next to the right as you clasp the leash, hands firmly planted against your belt buckle. At this point, don't worry about Una sitting or paying attention. Look out and imagine a square, sixty feet to a side. After saying, "Una, let's go," simply begin moving briskly in a straight line down one side of your imaginary square. Don't say anything more and look straight ahead, avoiding eye contact and letting the long leash drag along the ground.

There are any number of moves Una could attempt when you begin this, and with some dogs literally anything is possible. Since so many have not been properly leash-trained as pups and have never been restrained in any way early in life, many have an initial abhorrence of a leash. To some dogs, it is a tether that is meant to be chewed and broken; to others, it is a signal to run to a corner, curl up, and die. Other dogs might respond by attempting to melt into the ground or by scaling the handler's back and sides or by going into the "mule act" of refusing to move. However, usually the reaction is not so extreme. What most likely will happen is that Una will receive a sharp tug and begin following after you. At the tug, praise her briefly with "that's my girl!" She may run beyond you, around you, behind you. . . . No problem. Simply keep walking and stop after sixty feet; turn and face to the right. Una will likely be off to the side, waiting. Ignore her, and wait several minutes. Then begin again. Repeat this sequence as you go around the square. What you will soon notice is that Una begins to start walking along with you and that the slackened leash will be dragging along the ground. After fifteen or twenty minutes, she will seem focused and calm. You can treat and praise her at the very end of the session. When you repeat this exercise over several days (cutting down steadily on the length of time you pause at each corner), Una will be conditioned to walk along with you without fighting the leash, and you will be in a position to move on to the next stage of controlled walking.

For this, begin working with your six-foot leash. To hold the leash correctly, put your right thumb through the loop of the leash so that the leash lies across the open palm of your right hand. Make a fist with your right hand and lift the first two fingers so that they can grasp the leash a quarter of the way down. Rest your right hand on your right thigh. Grasp the other end of the leash with your left hand about two-thirds of the way down, knuckles forward. Rest your left hand on your left thigh, with Una to your left. This is your starting position.

Begin by saying, "Una, let's go," and immediately move out, giving her warm encouragement. The object here is for Una to walk along with you with a slackened leash, but without the precision of a formal heel. Make sure her collar is high up on the neck when you begin. If she starts to forge off to the side or out in front, simply give a light leash correction (a quick leash pop with your left hand) as you say, "Nah, let's go," and move off, either to the right (a 90-degree turn) or in a full-about turn (a 180-degree turn). Slap your left thigh with your left hand as you turn, to help direct Una's attention, and praise her warmly. The first couple of times you pop the leash, you may elicit a slight yelp, but as she responds successfully walking alongside you, praise and reward her. Don't stop to wait for her if she stops. Keep walking.

*With the dog on your left and making eye contact, you can begin the heel on leash. The dog's position does not have to be perfect in the beginning.*

When working with your dog on the heel, keep the leash loose, with your hands as illustrated in case you need to make a correction.

You will discover that changing directions is the most effective way to deal with inattention, with problems of forging and the like. By accompanying such change of direction with precise leash pops and varying your pace, you will get Una's undivided attention, and soon she'll be easy to walk informally, respecting the leash and needing only an occasional gentle reminder. If this is not the case, then we suggest using a prong collar to provide you with an added edge.

You can use this command ("let's go") when you want to allow Una the possibility of having a bit more freedom during a walk, so as to be able to sniff and investigate. You can expect to work at this level for about a week.

Walking in a controlled manner with a loose leash differs from heel in the manner of attention required of your dog. In controlled walking, as long as Una is not pulling, she is free to walk in a leisurely way, and not necessarily on the left-hand side. She can sniff, look around, interact with her environment. This is great for walks in the park or when she has to do her business. Once we move to heel, our expectations change. Now Una will be required to walk on your left side, paying strict attention to you while avoiding any tendency to forge, lag, crowd, or move off to the side. You will use the heel whenever you need strict control over Una, whether on a busy sidewalk or at a competitive obedience event. That said, let us repeat that the heel and sit introduced here are geared to companionship, not to achieving the level of precision demanded for competitive obedience. For most general dog owners, such precision would be a bit artificial.

## Formal Heeling

In teaching the formal heel, we assume that Una has gone through the preliminary exercises of leash training and controlled walking and that she has already learned the sit command in a general way from puppy training.* What we are going to do now is begin to heel formally and then attach a sit when we come to a stop. In time, this will become automatic; therefore, it is called the automatic sit. Begin with Una in the starting position. After making eye contact with her, step

---

*If this is not the case, see pp. 239–40 for instructions on teaching an older dog the stationary sit.

forward decisively with your left foot as you say, "Una, *heel,*" and start walking at a brisk pace. We introduce all active commands* with Una's name (since it is an effective attention-getter) but emphasize the command. Pronounce it clearly and audibly, and though your tone should be pleasant, don't baby-talk her. You want to communicate a certain quality of firmness and authority.

As you walk, be aware of where Una is. In heel, the goal is to have Una walking with her right shoulder adjacent to your left leg, paying attention in such a manner that should you change your pace, she would adjust immediately. Even having gone through the preliminary exercises of controlled walking, you can expect a certain amount of getting in front or lagging behind. Whatever Una's response, you can correct it if you know how to use the training collar properly while heeling. As you give a quick, sharp leash pop with your left hand, the training collar should tighten and release instantaneously, giving Una a helpful correction. We're not talking about taking her head off. The correction is simply a quick, attention-getting leash pop that occurs as the left hand slackens the leash, "pops," then immediately releases. Una will respond right away, perhaps even with a yelp of surprise, and then come back immediately to your side. Praise her instantly. Keep walking briskly, with plenty of encouragement, and start making frequent turns. For the time being, whenever you turn, repeat the command and offer her a treat every so often as you walk (from your left hand palm facing to the rear) to keep her attentive and focused.**

Remember: Una should feel your leash pops. Make your corrections fast and release the pressure of the collar immediately (if she is forging, you will most likely have to step into the correction with your left foot first to secure the necessary leash slackness to avoid pulling on the leash). Don't hang or haul the dog, and don't apply continual pressure on the training collar. Those methods (if you could even call them by that term) not only don't work but are inhumane. The "pop-snap" type of training-collar correction does work. It gets Una's attention and serves as a directional signal that is convincing.

---

*An active command is one in which the dog is doing something, moving into some exercise. In basic obedience, stand and stay are inactive commands.

**See *Raising Your Dog with the Monks of New Skete*. Heeling is covered in depth in tape three, *Obedience*.

*A graceful example of heel. The dog walks at the handler's pace. Keep the leash relaxed.*

People ask us about communicating with a dog during heel. For dogs that will not be competing in the obedience ring (where anything other than the initial command is not allowed), we suggest that you really open up and talk to your dog when you begin to heel. If your dog is coming along with you more or less peacefully, just whisper gentle encouragement, bending over slightly, making eye contact whenever you can. But if your dog forges ahead or hits the end of the leash, give a quick "nah" as you pop the dog back into place sharply, following up immediately with "good boy!" or "that's my girl" as the dog comes back into position. Be sure to make this exclamation very loud and happy. The dog will be momentarily disturbed by the negative correction coming from your leash, so your positive verbal correction must be loud enough to balance out things. What you are doing is providing negative and positive reinforcement in close succession, to achieve your objective of keeping your dog near your left leg. This means careful timing on your part. Under no circumstances should you verbally scold the dog for going out ahead or lagging. Use your voice positively in your heeling work. Keep the dog turned on, not off.

As Una heels nicely by your side, it is important to introduce her to turns. At this level of training there are three you will teach: a right-hand turn, a 180-degree turn, and a left-hand turn. Since Una is not at the point in her training where she is giving you her undivided attention, for now we'll preface a right-hand turn with an encouraging, suspense-filled "Okay, Una, we're going to turn now, ready? Okay, here we go . . ." just a few seconds beforehand, followed by "Una, heel" as you turn, then praising her as she follows. Doing so focuses her attention, letting her see that you are changing directions and giving her the chance to speed up and stay with you. If you like, you can use a treat to induce this positioning also; however, some dogs get a bit wired if they know you have a treat in your hand. If that's the case with Una, simply offer her encouragement with your voice. Follow this strategy for both right-hand and about-face turns. Once she gets used to turning, you can dispense with the wordy preface and simply stick with the "Una, heel" immediately before your turn.

To turn left, repeat the same sequence you used with the other turns, only now you'll need to slow down and draw your left hand back on the leash as you turn. Left-hand turns are particularly helpful for dogs who are not blatantly forging but are walking just a little out in front. A quick pop with the leash out over Una's head coupled with a "nah, heel," followed with praise and encouragement to watch you quickly teaches her to pay closer attention to you.

Now we are ready to add a sit, eventually refining it to the point where it becomes automatic whenever you come to a stop. Since Una has a basic understanding of sit from puppy classes, it can be taught naturally in conjunction with the heel. (If your dog has no background in sit and is starting out cold, consult the next chapter for instructions on teaching a stationary sit.) As you heel back and forth, begin to attach a sit by gliding into it off a right-hand turn. The momentum Una builds up going around you makes it much easier to glide naturally into the sit. As you come to a stop, transfer your leash completely to your right hand, pulling up on it slightly. At the same time, with your left hand reach down and back and gently press down on Una's rear end, easing her into a sit as you say, "Una, sit." Praise her warmly. You can also move through this sequence using a treat to induce the

*To teach the sit from a heel position, crouch down and gently touch the dog's rear while you pull up on the leash.*

sit, which will probably make the touch on her rear unnecessary. Now work on getting her to sit as you walk along a straight line. Do a series of consecutive heel-sit-heel repetitions. Once she seems to be getting the idea, eliminate the left-hand push down (or make the treats increasingly intermittent) but continue to lift the leash ever so slightly, praising her lavishly as she responds to your command.

Over the course of a week to ten days of daily practice, Una should be comfortable going into the sit whenever you give the command. Now, simply drop the command. Make the stop more deliberate by going into it with military-like precision and expect her to sit. If she does, warmly praise her (and/or give her a treat). If she does not, simply give a quick pop on the leash with "nah" and say, "Good girl," if she adjusts herself. Then immediately repeat the sequence. Over the course of several days Una will become increasingly conditioned to sit whenever you come to a stop. The key here is consistency. Once you begin working on the automatic sit, be resolved that your dog will sit each time you halt. Don't compromise unless you really want the dog to remain standing. You will appreciate the effort you've put into it when you come to a curb of a busy street and your dog sits without command, or when you are juggling a baby or a bag of groceries in your right arm and need to walk your dog with only your left hand on the leash. Dogs that heel correctly and sit automatically are a joy to introduce to guests. All that need be added is the "shake hands" trick for a truly impressive canine greeting!

## Possible Problems

Particularly if you have not leash-trained your puppy early on, you may experience any number of problems when teaching heel, even if you have started in the gradual way we recommend. For example, some dogs respond to the early stages of heel by clutching their owner's left leg in fear. Others begin to play "top-toe" by placing a paw on the handler's left foot, ready to move out when the handler does. If you get either of these reactions, especially the first fearful response, take things a bit slower. Chances are that you've been too quick with your dog while failing to include enough verbal encouragement. Stop, praise the dog, take a break together. Remember, every dog is an individual, and no single method of teaching any exercise is

*You can correct any forging ahead by quickly popping the training collar and walking in the opposite direction. Constant repetition of making turns to the right and to the left teaches the dog to be on the alert for your direction.*

absolutely right for every dog. Whenever you are training, be aware of what your dog is communicating —"read" her responses. Is she comprehending what you are doing? What do her eyes look like? Is she looking up at you? Never train any longer than twenty minutes. Two or three sessions a day are fine, if separated by several hours.

On occasion, your dog may suddenly forge ahead unexpectedly toward some distraction. You can try any number of corrections. As your dog begins to move out after some distraction, or just by sheer whim, grasp the leash tightly, hold your arms close to your body, and turn sharply to the right. At the same time give a loud encouraging vocal correction, "Nah, this way, that's my girl!" or something to that effect. If your dog is large and takes you by surprise with a bolt, pass the leash behind you from the left and come to a complete stop. Proceed only when the dog is by your left side. Your left turns will no doubt be less precise than your right turns.

If you have a dog who clutches your leg or hugs too close on left turns, give the leash a sharp pop away from your body as you turn into her. This helps move the dog out away from your body, preventing a trampling accident. As you continue to practice turns, have someone else watch you work with your dog and point out areas that

*It is helpful and more interesting to work on the heel with a group moving in a circle. You can go in either direction, always with your dog on your left.*

need improvement. An alternative is to train in front of a large plate-glass window or a sliding glass door and observe how you and your dog move as a team. Don't try to train completely alone. Constantly check and recheck your training guide, or consult an instructor.

One woman who had her dog trained at New Skete returned one afternoon to show off her teamwork with her cocker spaniel. Following our instructions to conduct a twenty-minute training session daily, she had trained her dog to a near perfect heel — with the dog walking on her right side. We gently reminded her that the dog had initially been trained (correctly) to heel to the left of the handler and that this had been carefully explained in a demonstration and in literature accompanying our course. "Oh!" she exclaimed. "That's why I had so much trouble getting him to heel!"

Another heeling problem involves the dog who is the vacuum cleaner, who will not pick her nose up off the ground. First, try using an object of attraction or a treat to get her head up and attention focused. Do a number of heel-sit sequences that are short and get her used to walking with her head up. Another approach is to, while you are walking with a loose leash, simply give gentle pops up when the dog's nose hits the ground. If there is no improvement, you may need to try using a prong collar, which is much more convincing. Again, keep the pops gentle and be very encouraging when her head stays up. Start with quick heel-sit sequences. Finally, a last alternative is to work in an area where there are distractions in the distance. This keeps the dog's interest up, again getting her used to walking properly.

Leash biting is another common problem that can surface early on in leash training. In this situation the dog responds to correctional snaps by attacking the leash. Probably he was never properly exposed to a leash as a puppy. If your dog snaps at the leash, give a quick jerk with the slackened leash straight up so that the clip at the end of the leash knocks the dog's jaw. Most dogs let go quickly. When you begin again, walk at a normal pace with the leash loose. Another solution to leash grabbing is to decrease your physical corrections and increase your positive vocal reinforcement. Don't slacken off completely, but try to minimize your physical corrections and maximize your vocal encouragement, until the leash biting stops.

"The mule act" is another common response, in which the dog braces his front and possibly hind feet and refuses to budge. The solution:

keep going. Don't try to "talk it out" with the dog. Simply turn around, make eye contact for a moment, and announce, "Let's get going." Give the heel command again, tap your leg, and then walk. If you stop longer than a few seconds to have a heart-to-heart coaxing session, you are ultimately doing yourself and your dog a disservice. Dogs who stage the mule act are often leader-type dogs who are used to taking their supposed "masters" for walks. Remember, in all obedience exercises, you are reordering your leadership relationship with your dog; so, at some point you are bound to be at odds.

You must curb all of the above problems as they happen. Don't wait until the next session. Your first heeling session might go smoothly, or it might be chaotic. Prepare yourself before you begin by reading all, not just part, of this book, making sure you understand the instructions. Practice the leash pop with a cooperative human* to help you sharpen your technique before you work with a dog. Finally, approach your initial heeling session calmly. Meditate with deep-breathing exercises just before you begin. Try to "center" on your objective of walking as a team with your dog.

---

*See *The Art of Raising a Puppy,* p. 177.

# 33

# The Stay

## The Sit and Stay

If Una is accomplished at heeling and automatically sitting when you stop, you have already essentially mastered the sit. If this is not the case and you prefer to teach Una the sit first, follow this approach to teach her a stationary sit. First, have Una in a standing position to your left (both of you should be facing the same direction). As you crouch down beside her, put your right hand on Una's chest, above her forelegs. (If she is bouncy or restless, you can place your right hand under her collar at the back of her neck.) Now place your left hand on her upper shoulders or neck. Stroking your left hand smoothly down her back, continue over the tail and then tuck her into a sit by putting even pressure on her chest as you tuck her rear legs. Say, "Una, sit," as you do this, and praise her warmly. Una will be sitting on your left hand, and your right hand will keep her from moving. Hold that position for five seconds as you continue to praise her. Repeat this process at least five times in a row, over the course of two or three days' training.

Next you can use a treat to help wean her away from having to be placed in the sit. With Una standing next to you, hold the treat in your right hand in front of her nose. Raise the treat slightly at a forty-five-degree angle as you say, "Una, sit." As she sits, give her the treat and praise her. Do this several times. Following this procedure, Una will learn the sit in no time and you will be able to combine it with walking on leash as described in the previous chapter. However, if after working with her for several sessions, Una still has trouble obeying the sit command, then correct her using the "nah, sit" in conjunction with a crisper leash pop. Speed up the pace of your work a notch, and emphasize the command word *sit* a bit more. Barbara Woodhouse, the famous British trainer, got a lot of mileage out of emphasizing the *t* in *sit*. You can dispense with this emphasis as soon as Una becomes more consistent.

There is one hand signal and one word common to all trainers in teaching the sit-stay. As they give the stay command, they bring their right hand in a sweeping motion directly in front of the dog's eyes, halting just before the dog's nose without touching it. Try this blocking hand motion on yourself a few times by bringing your hand toward your own face in a quick, sweeping motion. Stop sharply just before you reach your face. Your hand should be open, but your fingers should be closed together. You will note that the effect is dramatic — stay put. Make sure that your stay hand signal is quick. New handlers sometimes give the command several times, hesitantly, and then give a halfhearted, wimpy hand signal with the palm of their hand barely open. The dog naturally breaks the stay.

*The hand signal for the stay. Have the leash taut and up as you give the command to stay.*

*The stay signal may be given with either hand.*

With your dog sitting on your left, give this hand signal with your right hand, and with your left hand hold the leash straight up, applying a little upward pressure. (Bunch up the lead before you give the hand signal so that there is only about six inches of extension, plus the three or four inches the snug training collar allows.) Let your hand remain before the dog's face for now, and your leash taut but not choking the dog. Give the stay command. (Don't preface it with Una's name, since it is not an active command.) As you say, "Stay," step out with your right foot first and turn to face the front of Una. Hold the stay for a few seconds, then return to her side and praise her (you may also treat). Lessen the amount of tension on the lead as you progress. Stay at this level until Una is holding the stay for fifteen to twenty-five seconds.

Now you can begin to lengthen your distance after giving the stay command; however, go slowly and patiently. One of the reasons many owners have difficulty with the sit-stay is that they drop the leash far too quickly in the process. At this stage you need to be in a

*To reinforce the sit-stay, make eye contact, give the hand signal, and keep the leash up high.*

*Continue holding the leash straight up as you begin to walk around the dog clockwise.*

*Use both hands to hold the leash straight up as you walk behind the dog.*

position to correct Una instantaneously if she breaks her command. After giving the stay command, swing out in front of her to leash length, reversing the position of your hands as you do so (presuming you are right-handed). With the leash draped through your right hand between the thumb and first finger, you are now in a position to correct Una if she breaks the stay. At the first sign of this simply step into her and raise the leash sharply with a "nah!" This move pins Una right to the spot where she was told to stay. Repeat the command and hold the position for increasingly longer periods of time, up to a minute. Timing is crucial here — and this means you must read your dog. You have to be ready to correct Una at the first sign of her breaking the stay, when the thought is just occurring to her.

In your initial sit-stays, keep your leash in hand, holding it loosely. Only later, much later, when you are more confident of your dog's staying power, can you drop the lead. But even then, attach it to a long rope. This allows you to correct Una quickly and effectively in the event of a break. During a sit-stay, sustain eye contact with the dog to keep encouraging her to look at you. Vary your movements. Go out ten feet, then go from side to side, then return to Una's side. If her attention wanders, whisper softly and dramatically, "Watch me, watch me," to refocus her attention on you.

*If you work on sit-stays in a group, do it in a circle with the handlers bunched in the center.*

Next, proof her on the stay. Start with simple distractions at first, at leash length. Sidestep lightly, or toss up a ball or rock. If you are working in a class or with friends, try working in a circle, with the handlers bunched up in the middle and the dogs on sit-stays on the edge of the circle, facing in. For an effective distraction, the handlers can then rotate once and return to their original positions. Dogs who are weak on the sit-stay should be placed between dogs who are solid in their staying power. This helps the weak dog learn to stay. For now, simply return to your dog after a sit-stay. Make sure that when you return, you circle around in back of the dog, ending so that she is on your left side. Break the sit-stay with an animated "okay" and give your dog praise if the exercise was performed correctly. When your sit-stays are solid, you can utilize them in training what is perhaps the most crucial of all obedience exercises, the recall.

## The Stand-Stay

There is another command known as the stand-stay that we have found invaluable. The object of this exercise is to keep your dog remaining still in a standing position. This simple control exercise comes in handy during grooming sessions, when drying off a dog after he gets wet or muddy, and when examining him for ticks and other potential health problems. The dog who knows how to stand in one place for several minutes relieves you of the chore of having to wrestle him into place for whatever type of control you may need to exercise. It is not just a "frill" exercise. Taking the time to teach it to our dogs has saved us from many a strained muscle and moment of consternation.

Begin with your dog sitting on your left side. Kneel beside him so that you are facing him. Slide two fingers of your right hand through the training collar, under the muzzle, so that your palm is facing down and the back of your hand is under his chin. Placing your left hand under his belly, say, "Stand," as you press your right hand forward, parallel with the ground, and gently pull on the collar. With some slight upward pressure from your left hand, your dog will move into the stand position. Tell him to stay, praise him, and wait ten seconds before releasing him with an "okay" and guiding him back into the sit. Repeat the procedure several times per session, gradually

*In teaching the stand-stay, say, "Stay," hold the collar with your right hand, and with your left hand keep the dog from sitting.*

*Slowly move backward as you lengthen the lead and give the hand signal to stay.*

*Another way to teach the stand-stay using two leashes.*

lengthening the duration of the stand. At this stage, keep your hands on your dog. Make your initial goal that of getting him to stand for a full minute after a week of practice.

The next phase involves removing your hand and eventually moving away from your dog while he remains standing in position. Begin with the preliminary positioning described above. When you sense that your dog is relaxed and unlikely to move, give a stay command as you remove first your left hand, then your right. If he does not move, praise him gently. Should he break the stay, give a slight "nah" and place your hands back to their former positions. Once he is back in position, repeat the stay and try again. Soon he will be able to hold the stand-stay unassisted by you, and you will be able to move away from him as he stands in position.

To this point, the stand-stay has been executed from a stationary position. It is now possible to teach it while your dog is moving. However, be aware that some people experience difficulty with this transition as their dog tends to move into a sit by virtue of habit. Rather than fighting this tendency with potentially confusing corrections, we have found it helpful to use a second leash to assist the dog who is learning the distinction between stand and sit. Simply take a thin leather leash and loop it under his belly, holding the clip end above him with your left hand. With your normal leash in your right hand, walk at heel for several paces, holding the second leash with your left hand. As you come to a stop, say, "Stand," making sure to keep the second leash taut so that your dog doesn't move into a sit. Praise your dog warmly. As your dog gets the hang of things, start alternating between stand and sit. This way, you teach your dog the distinction between the two exercises, allowing you to wean your dog away from the second leash.

# 34

# The Recall

Most dog owners want two things from training. They want their dog to come when called and to lie down when asked. They may perceive other training exercises as ornamental and unconnected with their dog's ability to come when called and to lie down and stay. But there can be no real recall unless there is a sit-stay to practice it from, and no real lying down unless there is a down-stay. It is on the recall that owners of leader-type dogs experience difficulty.

Since dog training must be approached within the context of the overall dog-owner relationship, don't expect the techniques in this chapter to guarantee perfect performance in your dog unless you correct the other defective aspects of your relationship at the same time. For instance, don't expect the dog to run to you happily in a formal training session if you persist in calling him to you for punishment when he is off lead. If you have ever in the past called your dog to you and then punished him, resolve now to avoid doing so at all costs. Never call your dog to you for a correction. As a last resort, always go get the dog if you must reprimand him. But before things get to that point, think preventatively. Keep him on leash. Don't put him in a position where he is free not to come.

Begin to practice the come while you are heeling. As you heel, step back suddenly three or four paces and call Una in to you. Una will be surprised at this interruption of the heeling pattern and may continue to forge ahead. As she hits the end of the leash, give a quick pop that reorients her back to you, saying, "Una, come!" As Una nears you, pull up gently on the leash, as you did in training for the automatic sit, and have her sit in front of you. Now bend over and praise her.

When Una is coming in well and sitting automatically in front of you, begin to give the command and hand signal for stay, then walk around her and back to position with Una on your left side. Then repeat the routine several times. When you are sure she has the idea, put her on the sit-stay and move out farther. Do so incrementally, holding the leash taut over her head at first, until you can move out to leash length with the leash slack. If she starts to break her stay, simply correct upward with a "nah" and reinforce the stay. Later, move out to the end of the lead. Before proceeding, make sure that Una can hold the sit-stay for at least thirty seconds.

You are now ready to begin a more formal recall. We teach the command in three steps: leash length, trotting backward, and long-distance recall. The manner in which you take this sequence is very important and will likely influence your eventual ability to call Una reliably to you in off-leash situations. Above all, you have to keep this exercise happy and pleasant. The more natural and inviting you can keep your demeanor, the better. As we have said, our approach is geared toward the everyday dog owner, so we feel free to make use of voice and body language to aid in teaching this exercise, something not allowed in competitive obedience. Given that so many dogs are poor at this skill, the more inviting we can make the exercise, the better. Let the obedience aficionados work for the precision necessary for the ring — we feel free to use whatever gets Una coming consistently.

For example, how often have you seen a frustrated dog owner standing on her porch, hands on hips, stern expression on her face, and in an even sterner tone of voice repeatedly calling her wayward dog? Meanwhile, assuming that the dog even hears the command, she may be bounding about playfully, blissfully ignoring her owner. Finally, she might eventually crawl back to her master, fully expecting a trouncing. Many a dog owner pronounces at this point, "See, she knows she's done wrong — just look at the way she comes to me!"

If these dog owners could see themselves calling their dogs, they would realize that no living, feeling being, canine or human, would want to come to a person with that kind of body language and vocal expression. Not surprisingly, we have found that some handlers who train like Puritans and call their dogs like army sergeants often raise and call their children in the same manner. If you have experienced chronic difficulties in getting your dog to come to you, look closely at

*Phase one of teaching the recall. From the end of a six-foot leash, call your dog to you. Finish with the dog in a sit position and reward with praise and/or a treat.*

how inviting a target you have made yourself. Keep the end in mind. Realize that your tone of voice (elevated and happy) and physical bearing (open-armed and crouching) are extremely helpful in getting a consistent recall. Obviously, you can use treats as further motivators at each stage, knowing that you will wean your dog away from them as she gets better.

For phase one, with Una facing you on a sit-stay at leash length, crouch down and open your arms wide, calling her happily, "Una, come!" Chances are that your body language and tone of voice will encourage Una to begin coming. However, if she remains still, simply give a slight tug on the leash to encourage her (immediately letting the leash go slack), then guide her into a sit as she reaches you. Reward her with a treat and plenty of praise. Repeat this procedure several times. Over the course of two or three lessons, Una will grow quite comfortable with it. If you have to correct at all, make it ever so slight. Since you don't wish to force the dog, don't reel her in like a fish. Simply give a light tug and then release. You want Una to think it was her idea to come.

Now you can proceed to phase two. From the same starting position, open your arms as you say, "Una, come," and then start trotting backward. Una will follow after you. The purpose of this phase is to teach Una that she always has to keep the person calling her in front of her as her goal, thus avoiding "roadrunner" scenarios in which she merrily races past you. Though Una (and for that matter, any dog) is much faster than you are, you can anticipate her running past you to initiate the game. As she starts to veer away and run by, quickly reverse your direction, giving her a leash pop and a "nah," and continue trotting backward, changing directions and encouraging her as she follows. You will find yourself moving backward in a variety of patterns to keep Una focused on you. This is fine. Once Una catches on and is following you in a controlled way, bring her into a sit in front of you and offer her plenty of praise.

Now you are ready for phase three. Attach the clip at the end of your fifty-foot clothesline to the loop on your leash. Put a weight on the other end of the rope, so you can toss it out away from you or to another person more easily. Put Una on a sit-stay and toss the rope. Slowly walk backward ten feet, holding your hand out in front of you to reinforce her stay. As you face her, go down on one knee (eye level

*Phase two of teaching the recall. Begin the moving come on leash, trotting backward, with the dog following after you.*

Use your long line to practice distance recalls. Coil the rope in your right hand and hold the dog by the collar with your left hand.

Give the stay command as you continue to hold the dog by the collar and throw the coiled rope out in front.

*Walk to the end of the long line while the dog is on the sit-stay. Crouch down and give the come command as you throw out your arms in welcome.*

with the dog) and open your arms wide, creating a funnel effect, inviting her. As you open your arms, call her in a pleasant, enthusiastic voice. Use both Una's name and the word *come*. Clap your hands if you wish. As Una nears you, rise slightly but stay near the ground. Immediately after you say, "Come," smile broadly and try to make eye contact with her. Invite Una into your arms and give her a warm welcome as you guide her into a sit in front of you. If Una does not respond when you call her, simply give a light tug on the rope to get her moving and talk enthusiastically, being very encouraging.

Vary the length of time you wait until you call the dog to break the sit-stay and come in. Don't tolerate breaks before you call. If Una anticipates your calling her before you actually have, go get her and take her back to her original starting position. Next time you call, "Come!" chances are that your crouched position will encourage her to waddle into a sit as she nears you. If your dog is so ecstatic that she jumps up on you, allow it at first, in order not to dampen the dog's enthusiasm. Later, gently ease her down into a sit. The main point is to get your dog to respond willingly to the word *come*. Every time she does, the dog wins a big victory. Problems like jumping up, urinating from excitement, or happy rolling around on the grass can all be solved later. You are headed toward the goal of having your dog come in happily and sit in front of you, as is required in the obedience ring and is useful in home situations. But at first, it's acceptable to allow Una's imprecision when coming in. Again, the moment she enters the confines of your arms is very important. Let her enjoy it.

As you progress in recall work, lengthen the distance of your recalls, gradually going to the end of the fifty-foot rope. Also, don't call her immediately. Pace sideways, back and forth, so that she has to wait for your call. If she constantly breaks her stay, go back to shorter recalls and then work up to longer ones. Whenever you call her, give the rope a quick tug if Una shows the slightest hesitation. Praise her exuberantly when she reaches you.

When you are confident that Una fully understands the word *come*, begin to add distractions. Get an assistant to put a number of objects in her path, such as a bone, a ball, or another pet. Correct her interest in these diversions with a quick tug on the long rope. Don't be afraid to repeat the command "come" more than once (i.e., after the tug) and to use other orienting sounds, such as hand clapping, pounding the ground, or jingling a set of keys. Any supports that help are fine. But if Una's mind is wandering, don't waste any time popping the rope to bring her in.

Having your dog come and sit in front of you depends both on how well you have made eye contact with the dog when she heard you call and on your reaction to the dog when she is in front of your crouched body. From a distance, using a treat is the quickest way to induce a straight sit on the recall. As she approaches, simply raise the treat above her head and she will flow right into a sit. If you are not using treats, you'll likely need to be more patient. In the beginning, you can allow some playfulness, but begin to encourage Una to sit in front of you as you stroke her head and shoulders. Give the sit command if necessary, but try to ease her into a sit by upward pressure on the training collar and by petting the head area. We do not want Una to be forever dependent on the command "sit" when she comes in. We want her to glide into one on her own initiative.

When she is sitting in front of you consistently, continue to "funnel" her into you with your open arms, but as she nears you, rise to your full height slowly. When Una nears you and sits, give the stay command and hold the pose for a moment. This is excellent preparation if you intend to teach your dog the "finish," in which the dog returns to your left side by walking around your body or by sidehopping back to the left. This exercise is included in books that specialize in obedience ring exercises. The average dog owner might consider the finish ornamental, but it is a nifty exercise that can be very beautiful if performed correctly. Several of the books in the Select Reading List describe reliable ways of teaching it.

People often ask us, "How do I know my dog has a reliable off-leash recall?" There are several things to consider. First, for most every dog we have met, even those who have been titled and superbly trained, there is always *some* distraction that will be the one to cause them to err. No dog is perfect, so it is always wise to think preventatively. If you are going to be in an area with all sorts of enticing distractions, don't presume that Una will automatically come when you call, even when her recall seems flawless in most situations. Use common sense. Always have a backup plan in case Una decides not to come. Obviously, an electric collar can be used as good insurance once Una is well trained, but they are expensive and need to be used wisely.

For most owners, setting up tests in locations that simulate bona fide off-leash situations is the best way to determine how reliable the recall would be without a leash or rope — say, on a walk in the country. Make use of soccer or football fields at the local high school once you feel Una has achieved good consistency in the long recall. These are large areas that give the dog the feel of being off lead but are often enclosed by some sort of fence. Thus, there is little chance of Una getting distracted and inadvertently running out in front of a car, for example. Let her run around and have fun. Then, wait for a moment when she seems to be preoccupied with something else. If you call and she responds immediately, consistently, you'll have a good sense of her trustworthiness in most situations. If she starts to play games, go back to working on the long-distance recall with the rope. Always, always remember: don't let your dog off leash prematurely, until you are confident of her ability to come in 95 percent of the situations. If there is one mistake novice owners make it is this, and the consequences can be serious.

# 35

# The Down

We recommend any one of three methods when teaching the down. The first two involve easing Una into the down position rather than forcing her down, and presume some body contact. The third method induces Una into a down with the help of a treat, and is a simple extension of the puppy down. Many people have great difficulty training the down, which is actually a simple exercise. It is often approached as a punitive exercise, with irate handlers attempting to ground their dogs by stepping on the leash or strangling them until they lie down. This frustrates both handler and dog.

The following methods of training the down are humane and easy, and most important, they work. Begin by heeling a bit with Una, then come to a stop with her in a sit position. Make eye contact by stroking the right side of her face, which encourages her to look up at you. Mention calmly, "Okay, ready for something new?" Make your voice animated and happy. Many handlers become tense when they begin to teach the down, anticipating that they may have trouble and thereby actually inviting it in the process. Immediately tell Una to stay, and kneel down beside her. With your right hand grasp one leg just below the elbow, and place your left hand on her upper back. Give the command "down!" in a firm but animated voice and move the leg forward, simultaneously pushing down firmly on the upper back. Do not make the motion abrupt and coarse — make it smooth, easing the dog down. If you need to, practice this exercise with a human on all fours before attempting it with your dog.

Once Una is down, kneel nearby, stroke her, and give her some reassuring praise. Keep your tone of voice low but affectionate, or she will try to get up. Keep one hand on the back region, applying light

*Teaching the assisted down. While your dog is in the sit position, lift his front paw as you give the "down" command, slowly lean over and down on him, and thus gently ease him all the way into the down. Then give the "stay" command and hand signal. Finally, give lots of praise to your dog and to yourself! Continue to repeat this assisted down until the dog responds to the "down" command without help. Praise all around again.*

pressure. In the early stages, end the exercise after thirty seconds or a minute, praising Una when she rises. Decide on some key word like "okay!" to signal her to get up. When she seems to be steady on the down, you can begin to rise from the ground yourself, but keep one hand in contact with her back.

Next place your left foot over the leash near where it is attached to Una. This keeps Una in place and helps her get comfortable with the position without your having to be right beside her or to correct her repeatedly. Soon you will be able to take your foot off the leash. Correct her attempts to get up by applying pressure on her back and saying, "No, down!" and placing her back in the down. If she moves quickly and manages to squirm up, go through the whole process again, then rise more slowly, applying pressure as you get up. If she tries to move backward as you lower her front, block this motion with your left leg or knee placed squarely behind her rear end. Don't be afraid to use all parts of your body in training. Your legs can become auxiliary arms if you know how to use them.

Some larger breeds (German shepherd dogs, Great Danes, Saint Bernards) might need even more body contact and pressure to ensure a smooth down. For these breeds, or for dogs who fight the first method, try a slightly modified approach as follows: lift both front legs together, with your hands just below the dog's elbows, and lower her front. You can either drape your left arm over her back to clasp her left foreleg or insert your right hand and forearm behind Una's two front legs (as you hold her collar on the back of her neck with your left hand) and then lift the front legs and lower her to the ground, saying, "Una, down!" At the same time, you may need to lean over and

*Another method of teaching the down is to lift both front legs and lean over and down on the dog, easing him into the down as you say, "Down."*

into her, leaning your left knee on her back. Follow this move with ample praise. The first few times, you may find this movement unco-ordinated, but with repetition it becomes smoother. Since your face will be quite close to Una's, this method presumes that Una is not aggressive in any way. Initially, your goal is simply to get Una used to being placed in a down position. Once you have her down, proceed as above. With some particularly rough types who refuse to stay down for any length of time, simply place your foot on the leash, which prevents them from getting up. If the dog throws a tantrum of sorts (something that could happen if he was not introduced to down as a pup), you may have to crouch down beside him and proceed more patiently. But don't lose your cool. Keep calm and proceed step-by-step, as you are able.

If you are able to stand up for three minutes and the dog stays down, consider the exercise partially learned, and begin to teach the release, which is to sit. To end the exercise, say, "Okay!" and slap the upper part of your thigh, encouraging the dog to rise. Stationed on the dog's right, you can block any attempt to come to a full stand by having your left hand aimed toward the dog's rear end, ready to push it down.

The third method of teaching the down is good for sensitive dogs, as well as those who have had no previous exposure to this exercise. Crouch in front of Una and place your left hand in her collar, under her muzzle. Show her a treat with your right hand. In a downward movement, induce her into a down position, putting very light pressure on the collar as you say, "Una, down," and guide her into the down. Give her the treat and soft praise. Repeat this sequence five times. Next, instead of starting out in front of her, crouch beside her. Put your left hand on Una's withers as your right hand shows her the treat at eye level out in front of her. As you command, "Una, down!" move the treat toward the ground as you apply light pressure on her withers with your left hand. As she moves into the down, let her have the treat and praise her warmly.

When you complete a down, no matter what its length, be sure to praise the dog lavishly. Stroke especially the side of the face closest to your body to encourage her to look up at you. Make eye contact and sustain your verbal praise for a while.

There is a potential problem that may develop when training the down. Una may come to depend on your manipulating her into position, thinking that the point of the exercise is to wait to be lowered. For this type of dog, stop lowering her. With Una in a sit position, stand next to her (facing forward) and put your left hand where the collar and leash join, just under the ears. Now fold the leash entirely in your left hand. Turn sideways so that you are facing Una, just a bit in front of her. Now, say, "Una, down," as your right hand passes downward in front of her face. As you do so, apply steady pressure with your left hand and guide Una into a down. You may need to tap the ground several times with your right hand, encouraging her as you go. Again, once she is down, keep one foot over the leash to prevent her from rising.

# 36

# The Down-Stay

If Una really knows the down, then the down-stay should follow naturally. Most dogs do not like to lie down and then immediately get up. When they settle down, they mean to stay down for some time. The practice of circling two or more times before "landing" is common to many breeds, especially German shepherd dogs. The dog appears to be carefully selecting and surveying the eventual resting place. If you are training for the obedience ring, you have to eliminate this practice by teaching an immediate down and insisting on an unwavering down-stay. A sloppy down or fidgeting on the down-stay costs points in the obedience ring. If you are not training for the ring, you may decide to permit quirks like circling before lying down, lolling about or rolling over on a down-stay, or extensive yawning ten seconds into a down-stay. Una is not bored. It's a position that encourages rest. Nevertheless, don't let her become too casual. If she starts prolonging the "pre-down" rituals, simply correct her and speed up expectations.

To begin work on the down-stay, after Una has gone into the down, give her the command "stay" along with the hand signal, as in your work for the sit-stay. Stand next to her until she can stay for a full minute. Now start varying things. Circle Una while keeping one hand in contact with her back. Circle twice and release your hand, but do not allow her to get up. Repeat this circling process, applying pressure with only the tip of your index finger. You will be stooping over as you circle her. She may try to look behind at you as you circle. Allow this, but don't permit any attempts to get up. Curb any such attempt immediately by giving her a quick leash pop toward the ground as

*Stay in sight while your dog does the down-stay. Avoid eye contact now by looking to the side. Otherwise, the dog will think you are inviting him to come.*

you say, "Nah!" following up with "down, stay!" When you feel confident, go around Una without stooping down or applying hand pressure. Don't be afraid to repeat the command "stay" several times if she shows any interest in rising.

There is nothing wrong with her looking around at you as you circle and go momentarily out of sight. But don't encourage this trait by talking to her from the rear. Once they understand what is going on and what is expected of them, most dogs simply look ahead and wait while the handler walks around. When your dog gives you this type of reaction and no longer appears worried or jittery about staying in one place, you can begin adding distractions to make sure the down-stay is steady. Begin by stepping over the dog, back and forth, several times. Throw a ball or twig in front of her, repeating the command

*With your dog in a down-stay, jump over him several times back and forth so that he learns to stay put.*

*Nothing surpasses working in a class for steady stays. Here, as a distraction, the monks and nuns toss volleyballs down the line of handlers while the dogs hold a down-stay.*

*An agile handler can sprint over dogs while they hold the down-stay. Distractions can strengthen all areas of training.*

"stay" as you throw the object. If you are working in a class, an agile handler can leap over each dog in line. In a class, the dogs learn from one another not to move. Most of our puppies at New Skete learn to lie down and stay at a young age without any formal training simply by watching their mothers and older dogs on down-stays when the monks are dining. A dog who is just learning the down-stay sometimes does better if sandwiched between two experienced dogs who are steady on the exercise and distraction-proof. These self-assured dogs help allay the fears of the novice dog and help him hold the down-stay longer.

The correction for breaking the down-stay is always the same. Go get Una. March her back to the same spot and place her on the down again. Hopefully, you will be able to stop her in her tracks by a sharp "nah!" which may pin Una back to the ground before she moves more than a step or two. If she lies back down on her own initiative, it is not necessary to replace her in the same spot.

Next to the recall, the down-stay is probably the most useful of all the exercises in formal obedience training. Begin to integrate this command into Una's daily routine. Have her lie down and stay while you eat dinner, read, or watch television or during any sustained activity. Build up Una's ability to do so by staging practice sessions during coffee breaks. This way, if Una breaks several times, you'll feel prepared to get up and correct her, as opposed to dinnertime, when you might be tired and when continual corrections may mean that your dinner gets cold. The more Una can learn to see herself in a down while you are at table, the longer and more reliable the down-stay will become. When company visits, allow her to greet your friends, then place her on a down-stay nearby. Be consistent on correcting any breaks. Simply get her and put her back in position. Also, don't expect immediately to isolate her from you and expect her to hold a down-stay. This takes a lot of practice. If you plan to go on to CDX (Companion Dog Excellent) level in the obedience ring, your dog needs to be able to hold a down-stay with a group of dogs for five minutes while you are out of sight. If you are building up to this level of training, you are most probably also working with the high jump. If you occasionally hide behind the high jump while your dog is on a down-stay, the dog will never be sure whether you are behind it at an actual obedience trial.

In a household situation, the down-stay can be used for dogs who bother company or display aggressive reactions to humans or other dogs. Practice training your dog to the point where the down-stay is perfect. Dogs who know and can sustain the down and down-stay rarely need to be shunted off for any reason. They can be told to lie down and stay, and are able to be peacefully included in the family circle. This is a distinct pleasure for both dog and owner.

# 37

# About Obedience Competition

If you really want to see what dogs working with their owners are capable of, attend an obedience trial. If you and your dog are proficient in obedience work, you may even consider entering the obedience ring yourselves. These events differ from conformation shows, which emphasize the structure and beauty of the dog. The American Kennel Club offers a Companion Dog (CD) title, Companion Dog Excellent (CDX), Utility Dog (UD), and a Tracking title (T). There is now also an Obedience Trial Champion title (OTC). A booklet describing the obedience regulations can be obtained from the American Kennel Club website (www.akc.org) or by writing to 51 Madison Avenue, New York, NY 10010. Since these regulations sometimes change, send for the latest edition.

If you are seriously interested in obedience competition, you should subscribe to one or both of the periodicals specializing in obedience work (see Select Reading List). You should also join an obedience club. Try to find a club that specializes in obedience training; if that isn't possible, join the training branch of a local breed club. Obedience training, on the whole, provides an opportunity for mutual growth, on the human as well as canine level. We have met hundreds of people who enjoy the world of professional obedience competition. There is one catch: you and your dog must be good at it to get anywhere in professional competition.

If you are considering obedience competition, make sure that you and your dog are able to work as a team. Don't enter the ring until you know the exercises and are completely ready. If you jump in too quickly, you will probably fail the trial, waste the judges' and other entrants' time, and possibly do damage to your dog. Take your dog to fun matches before you go to a real obedience trial.

Puppies intended for obedience work should be purchased from reliable, behavior-oriented breeders who breed for brains. These youngsters should be exposed to KPT and taken to fun matches early in life. However, it is seldom too late to attempt to win the CD or a higher title. If you feel you are ring-ready and still do not qualify with the necessary 170 score, don't be discouraged — chances are, you mishandled your dog. Check with the judge when he or she has a free moment. Most judges will tell you what you did wrong, and many may give you helpful advice. Above all, don't become bitter or disappointed. These reactions have a way of affecting training. Your dog picks up on these reactions immediately, and they damage your overall performance. Bear in mind, too, that you are competing for points and not against other handlers and their dogs. If you want to compete in the obedience ring, you must have a sense of humor. You must be able to laugh at yourself and your dog. The first thing you will learn, sometimes painfully, is that your dog has some faults and that your dog is not the center of the universe and the idol of all. These two realizations alone are worth all the time and effort that goes into obedience competition.

Though this book is not specifically intended for obedience competitors as such, we feel that it can be useful in enabling them to "read" their dogs more effectively and to help them perfect the exercises involved in winning obedience titles.

The American Kennel Club puts the matter succinctly in its introduction to the obedience regulations: "The purpose of obedience trials is to demonstrate the usefulness of the pure-bred dog as a *companion of man* [italics ours], not merely the dog's ability to follow specified routines in the obedience ring." If you decide to enter obedience trials, keep your purpose and goal in mind.

# Problems

# 38

# Understanding Your Dog's Personality: The Problem Explained

An owner brought in an eight-month-old rottweiler named Sasha for our three-week obedience-training program. During the interview it became increasingly clear how dissatisfied the man was with his dog, and when we inquired about his efforts at training Sasha himself, he responded, "I've really tried, but I just can't figure her out. She's a nice enough dog around the house, but whenever we work on training, her ears go back, she lags and starts to pull this creepy-crawly routine, as if she were afraid of me. It's embarrassing. . . . She's a rottweiler, for goodness' sake! I've never hit her, yet from her reaction you'd think I'd been abusing her. It's so frustrating. She's totally different from the last rottie I had." When we asked him how he had tried correcting this submissive behavior, the man exclaimed, "I try to be consistent in correcting her with the leash, trying to get her to walk with me. Believe me, I correct and correct, but she doesn't seem to get the point." We did believe him, which is why we explained that in this case we were going to take a very deliberate, praise-oriented approach that would also use food to work with Sasha. The man seemed surprised. "But I thought you didn't use food in training. . . ." "Not necessarily, particularly when it's in the best interests of your dog," we replied without hesitation. Just because Sasha was a rottweiler, a breed that is notoriously dominant and serious, doesn't mean that we could ignore her specific personality. We needed to adjust our training accordingly. Let us explain.

During the thirty years we have been training dogs of all breeds, one of the clearest lessons we have learned is that each dog is a unique creature. Despite the fact that selective breeding has resulted in a certain general predictability regarding breed characteristics, there are plenty of individual representatives who vary from the norm. Contrary to what one might ordinarily expect, we've trained extroverted shar-peis as well as bashful golden retrievers. No two dogs are the same, which is why we never grow bored with them and why we find them endlessly fascinating.

Good training should allow your dog's uniqueness to blossom in a way that enhances your relationship. However, to accomplish this goal, you have to be an educated and sensitive observer of your dog, recognizing that general training guidelines always have to be tailored and adapted to each particular dog and not always applied in the same way. Throughout your life with your dog, it is crucial to remember this basic principle: with respect to training, always take into account the dog at the end of the leash. Any specific training approach has to be adapted to the particular personality of the individual dog in question.

If we are going to be good trainers, we have to be flexible and continually learning. Rather than being captive to any one "system," we may have to become familiar with a number of different approaches and draw from that knowledge as circumstances require. We want to be able to focus on the needs and temperament of an individual dog, on what will most help this particular dog learn in this particular situation. Sometimes that may involve food inducement, at other times, much less so; sometimes a prong collar may be particularly helpful in training, in other cases it would be positively harmful.

Taking this approach toward the individual dog requires that owners understand something about their dog's personality. For a dog lacking in self-confidence — like Sasha, for example — food rewards and enthusiastic praise are valuable motivators to help her get out of herself and into a positive, self-confident demeanor with respect to the training. That might not be the case with an exuberant, happy-go-lucky standard poodle.

## Consider the Dog

Personality refers to the particular canine temperament that your dog manifests. Broadly understood, this includes genetically determined behavior as well as behavior that has been shaped by your dog's particular environment. For example, all dogs possess a number of instinctive behaviors, or drives, which strongly influence how they interact with the world, and in turn respond to training:*

- Prey drive, made up of behaviors associated with hunting, killing, and eating, manifests itself when your dog chases after a speedy squirrel in the backyard or vigorously shakes the play rope. A dog with a high degree of prey drive is challenging for an owner to train, since the dog is easily distracted by moving objects and may require a lot of work with set-up situations that teach him to focus on the handler.
- Pack drive involves social behaviors related to your dog's being part of a pack; it includes sexual and parental behavior and is also readily seen when your dog actively solicits play from the dog next door, for example, or when he shadows you around the house. Since a vital aspect of pack behavior is the willingness to work as part of a team, this is the drive you'll most frequently draw upon when teaching obedience exercises. A dog with a high pack drive is easier to train because she responds readily to touch and praise. She enjoys being with you and finds training stimulating and enjoyable.
- Defense drive includes those behaviors related to survival and self-preservation that are manifested in either fighting or fleeing, as when your dog scoots away from a loud noise or barks threateningly at the door after the doorbell has rung. Whereas a dog with strong fight instincts demands solid leadership skills and probably has little problem with firm corrections in training, a dog with a high defense drive oriented to flight needs to be watched carefully. You need to keep your training positive and upbeat, avoiding any force training and most likely using an elevated, pleasant tone of voice and food reinforcement.

Though all dogs possess these drives to varying degrees, what is important to be aware of is that each dog manifests them in different

---

*The information on drives in this section is highly indebted to the insights of Jack and Wendy Volhard, who have written at length about them in several of their books, most recently *The Complete Idiot's Guide to a Well-Trained Dog.*

ways, in different combinations. And this is not the end of it. Environmental experiences can either reinforce or modify these instinctive drives — for example, the crucial role of early socialization in puppyhood, the effect of traumatic experiences such as gunshots, fireworks, and car backfires, and the basic health or confusion of the owner-dog relationship. Now we can begin to appreciate the complex character of an individual dog's personality. Even though Sasha happened to be a rottweiler (a breed that traditionally manifests a high degree of protective/defensive drives), her particular personality expressed itself more neutrally, or even with what we would describe as a noticeable lack of self-confidence. Good training tries to deal with the dog as she is and then discover the best way to realize her potential within the bounds of her living situation.

So how can we apply this principle? In our discussion of puppy temperament testing in *The Art of Raising a Puppy*,* we describe the puppy test we use here at New Skete to aid us in placing our German shepherd pups in their new homes. From our experience using this test in connection with puppy placement and obedience potential, we are able to single out three aspects of canine personality relating to their drives that strongly influence how you go about training your dog: sociability, dominance, and touch sensitivity. All dogs manifest varying degrees of these traits. To train your dog intelligently and successfully, it is vital to have a clear sense of how your dog fits these three categories.

## Sociability, Dominance, Touch Sensitivity

1. Is your dog outgoing and people-oriented? Or perhaps she is somewhat reserved and aloof? Usually this is fairly easy to judge. The happy dog who eagerly solicits attention from houseguests by jumping all over them is clearly the people-oriented extrovert, whereas the independent, reserved dog may give a few barks and then simply retire off to the side and want little to do with the visitors. More extreme yet is the shy dog who shivers and shakes in the corner when faced with an unfamiliar face. Such a dog may even growl or bark in distress with his tail tucked.

---

*The Art of Raising a Puppy, pp. 62–70.

**2.** To get a clear idea of the broad range of possibilities in regard to canine dominance, think of the pack. In a pack you always have alpha (first) and omega (last) members, with the rest ranking somewhere in between. The dominant alpha exhibits dominant body posture: head-on greetings with strong eye contact and raised hackles (neck and back hairs). The tail is held high, with ears erect. Such a dog exudes confidence. At the opposite end of the spectrum is the highly submissive dog who greets other pack members by rolling on his back and submissively urinating. Such a dog is a placater, who will grovel to defuse any threat he perceives from a more dominant pack member.

**3.** Finally, dogs manifest various degrees of touch sensitivity. Some dogs are touch-insensitive, which means that they do not react as negatively to circumstances other dogs might find unpleasant. One of our German shepherds, for example, loved to go after porcupines — even after she had been quilled several times — and reveled in racing through our creek . . . even in freezing-cold temperatures! At the other extreme are touch-sensitive dogs, who are very sensitive to any discomfort whatsoever. Such a dog may balk at going out to relieve himself when it's raining, or whimper and cry at the slightest muscle strain and dislike being petted. Dogs in between, of a more normal sensitivity, respond evenly to leash pops and recover quickly from unexpected exertion or tumbles they might experience playing with another dog.

Given these realities, you as a responsible owner want to take into account the particular personality traits your dog exhibits and tailor your training methods accordingly. It is beyond the scope of this book to script all the possible variations in canine personality and the different ways to approach training, but we can give you some clues. These insights, combined with your own knowledge, intuition, and observation of your dog, will help you move in the right direction. By taking into account your dog's personality, you can reasonably anticipate how he will respond to training and can predict the type of demeanor appropriate for you so that you will be more effective. It will also help you judge the appropriateness of certain kinds of training equipment.

For example, just as it makes little sense to use heavy doses of compulsive training with a wallflower-type dog who withers merely at a piercing glance, it may be entirely necessary and appropriate to use a prong collar with a larger, more dominant dog who is touch-insensitive. With the former, food is particularly helpful in training. It is also crucial to praise and affirm only "brave" actions and to avoid any sort of sentimental, smothering type of behavior that inadvertently reinforces the fearful, submissive behavior. With the more dominant dog the trainer needs to manifest strong leadership and firm self-confidence. Such a dog responds well to crisp leash corrections and must never be allowed to lead. You'll need to hold his attention by keeping the pace of training quick and filled with praise.

The independent, aloof dog (often found in many of the Oriental breeds) is different still. Such a dog, often lacking in pack drive, can be extremely stubborn in training, making it difficult to force her to do something she doesn't want to do. Instead of declaring war with force, it is much better to induce her into doing what you want and then make her feel as though it were her idea. Food is a helpful motivator here. She needs to do an exercise a number of times before she feels comfortable with it, but then responds well.

Then in turn there is the "live wire," high in prey and pack instinct, who may find it difficult to focus on any one exercise since he is ready and looking to act on every distraction that comes his way. Aside from making sure this dog gets lots of exercise, you need to keep his training sessions short and positive; light, frequent leash corrections can help maintain his focus, and your calm demeanor helps the dog progressively to extend the length of his attention span.

Different still is the more even-tempered dog who relishes the opportunity to work and learn new things and who is on the lookout for the first sign of an upcoming training session. This dog responds well to inviting body language and warm praise; she handles light corrections easily and responds to the tone of your voice. Such a dog has the ability to make a novice trainer look like a pro simply because she is so connected to the training adventure — and forgiving of mistakes!

Can we say which sort of personality is best? In a word, no. The ideal canine personality is always an extremely subjective concept: it depends entirely on what your needs and desires are. Often it is determined by the work the dog has been bred and selected for. Though many trainers prefer working with dogs who have high pack drive and moderate levels of prey and defense drives since they respond so nicely to training, others enjoy the challenge of working with breeds that are not generally known for their trainability. We know one trainer who relishes training "hard" dogs (high in prey and defense drives, lower in pack) because of the spirit they show in advanced forms of protection training, yet who would be the first to point out how ill suited such a dog may be for a family with small kids living in suburbia. And we know still another dog owner who loves her three shih tzus simply for their capacity to bring comfort and pleasure to her but who has no illusions about how challenging they are to train in simple obedience. Though this book is geared especially for owners whose prime reason for having a dog is companionship and who will most likely be interested in higher levels of sociability and friendliness (with moderate to lower levels of prey and defense drives), we have tried to present our approach in a way that is helpful also to owners with dogs who fall into other, more challenging categories.

# 39

# House Training

There is no area of dog training where more myths abound than house training. House training should be a simple procedure, but for many owners it is a drudgery that sometimes never ends. It's surprising the number of clients we see who have owned dogs for over two years and confess, "He's really never been housebroken." Later we come to find that this can mean that the dog has the problem on a daily, and even twice-daily, basis. If house training had been approached correctly and consistently at first, the problem would have been short-lived.

**Myth Number 1:** When you find an accident, rub the dog's nose in it. Do *not* do this, ever. You defeat your own purpose, risk infecting the dog, and encourage stool consumption.

**Myth Number 2:** When you find a mess, you should hit the dog and then ostracize it for a good long period. Nonsense.

**Myth Number 3:** After you find a mess, take the dog to the place where elimination is supposed to occur, stand over him, and scold him. This is all backwards: the dog needs praise and encouragement at this location, not punishment.

**Myth Number 4:** You can train a dog to eliminate on papers or outside, or both. This myth causes more confusion than any other. Dogs need a consistent approach to house training. We never suggest paper training unless the situation absolutely demands it and the dog has no quick access to the outdoors. (For instance, because of the danger of disease, city dogs are often quarantined inside until they have had a full cycle of vaccinations, that is, about sixteen weeks of age. This demands paper training first, then house training, once your dog is able to go outside.) For owners without these specific circumstances, we find that paper training is usually an ill-conceived shortcut for the owner, a convenience at the time that later backfires if it is even successfully comprehended by the dog to begin with. If at all possible, train your dog to eliminate in one spot, outdoors.

Here is our approach to house-training and house-soiling incidents.

1. The basic rule is to capitalize on your dog's natural desire to keep her nest clean, an inherited characteristic. Follow the guidelines for crate training given in chapter 29. When she's not in the crate, have your dog with you, so as to monitor her. Anticipate when she wants to go out. The signs are nose grazing, obvious squatting, loitering around the door, constant activity. The times are every two hours for puppies younger than twelve weeks old (except when they are asleep), after waking from a nap, fifteen minutes to half an hour after eating, before riding in a car, after drinking a large amount of water.

2. We suggest regular feedings for pups and older dogs, unless modified by your veterinarian's advice. We find that the "nibbler plan" (food down all day for the dog to eat at will) often encourages house soiling because the dog is not on a predictable timetable. For pups we suggest that water be offered at regular intervals, not left down all the time. Water and food should be taken up at night, giving the pup (or older dog) enough time to eliminate before he retires. Do not leave water down for a pup until you are sure of absolute sphincter control.

3. Respect your dog's biological clock. Be consistent in the times you let her outdoors. To a degree, many human activities have to be coordinated with the dog's biological schedule until she is mature. Sleeping in, late parties, vacations, shopping trips, and other activities may have to be based on when the dog will need to go out.

4. If you discover a mess, don't call the dog to punish him. Dogs do not understand the connection between an after-the-fact correction

and the mess. All you accomplish by such a correction is to damage your relationship. To correct the dog, you must catch him in the act, using a physical correction for an adult dog and immediately taking him outside. For puppies, we discourage strong corrections — occasional accidents are part of raising a pup, and house training needs to be approached preventatively, by monitoring your pup's behavior and responding accordingly. If you observe your pup getting ready to eliminate, don't shriek or freak out — swiftly sweep him up and take him outdoors, even if this means an elevator trip. Most pups stop short and hold it until they get outside.

**5.** If it's too late for that, calmly take the dog out of the room without praise, and confine her to her crate while you clean up the mess out of her sight.

**6.** If you are just beginning house training, do not make a big thing over it but scold or growl to make the dog feel the effects of your displeasure. For older dogs with chronic problems, physical discipline may be in order unless they are ill, but again, only when they are caught in the act.

**7.** If you catch your dog in the very beginning of the act, correct him and take him briskly to the proper place and let the dog continue there. This is essential because the dog must make the connection on where to eliminate. Stay with the dog if it's not possible to leave him alone.

**8.** After you return, clean up the accident with paper towels or other absorbent materials. Paper plates cut in half make good emergency scoops. Wash the area with a 25 percent solution of white vinegar and hot water. Odor neutralizers, like Nilodor and Lysol, can be used after the vinegar treatment. It is important to remove the scent from both human and canine detection. If possible, block off the area until it is dry. An overturned chair will do. Again, do not let the dog see you clean up her mess. Seeing you clean it up encourages the "maid syndrome" in bossy pets, which is at the root of many serious house-soiling problems. You are not the maid in residence to clean up after the dog. Nor are you the doorman, there to let the dog out whenever she demands. Though house training means anticipating your dog's desires, it also means the dog must quickly learn sphincter control. You defeat this development if you play maid or doorman without taking further steps to house-train her. After a mistake, don't isolate the dog or keep scolding her. Avoid overfondling the pet at this time, unless in response to a command (come, sit, stay, lie down).

The final problem related to house training is marking behavior. This is a trait most commonly seen in male dogs who lift their legs and deposit a splash of urine on an object. (Leg lifting also occurs less frequently with some females; however, females usually mark directly on the ground to advertise their availability when they are in season.) Behaviorists believe that in addition to marking territory, urine also provides vital information about a dog (i.e., gender, age, and rank). Male dogs lift their legs so that their scent can be placed higher than that of previous markings, and the frequency with which it occurs on walks can be nothing short of amazing! The behavior is normal and natural but can be inconvenient or embarrassing in certain circumstances. For example, you take Duke over to a friend's house (especially someone with another dog) and suddenly notice him dousing a piece of furniture. When your family moves into a new (or newly decorated) house, or when there is more than one other animal in the house, marking is prone to occur.

When marking is an ingrained problem, neutering is in order. In addition, however, close monitoring is crucial. When at a friend's house, don't let Duke out of your sight and be ready to take him outside at the first sign of inclination. If you catch Duke in the act, correct him and take him immediately outside. Otherwise, keep him on a down-stay.

If you are experiencing more pronounced difficulties in house-training a puppy or older dog, seek advice from a qualified trainer or behaviorist who will sit down with you and map out an approach for your dog. Avoid harsh physicality in disciplining house soiling; be alert, consistent, and think preventatively.

## 40

# Chewing, Digging, and Jumping Up

## Chewing

If breeders did a better job of preventing early chewing, their puppies probably would have fewer problems when they go to new homes. As with many canine troubles, the problem starts at an early age. Some breeders may not provide chew toys, instead allowing their puppies to chew on portions of the whelping box or kennel. All it takes to stop this kind of behavior is a loud "no!" and replacing the object being chewed with an acceptable one. When breeders provide hanging and toss toys for the litter, it helps puppies focus on appropriate objects to chew. In general, when the puppy goes to a new home, the owner should focus the pup on a few preferred toys only (Nylabone, rope toy, and Kong are good). Keep one particular toy in each location where the pup spends time (i.e., bedroom, living room, kitchen, crate). Don't go overboard: if the pup has too many things to chew on, the message he receives is that everything is chewable.

Puppies need to chew — chewing helps baby teeth to loosen, making room for the permanent teeth to emerge. Your job is to focus puppies' oral attention on appropriate things. New puppy owners should be aware of the teething period, which is especially pronounced between four and six months of age. A good do-it-yourself toy for teething pain is a frozen washcloth that has been twisted. The coldness helps numb the pain. When the washcloth thaws, simply wash it thoroughly, twist it again, and stick it back in the freezer to be used again. Troubles connected with puppy chewing are explained further in chapter 29, on puppy training.

Several commercial products are available to help puppies and older dogs stop chewing. They work on the premise that some dogs withdraw from unpleasant smells such as citronella, Tabasco, or even Listerine. A new repellent containing methyl nonyl ketone is effective. Grannick's Bitter Apple lotion is an old standby of experienced trainers and breeders. But don't count on sprays, ointments, or magic salves to relieve chewing problems. Though they may help to a degree, the best method of chewing control is early vigilance, a timely reprimand, and a refocusing of attention on an appropriate object or toy.

The length of time destructive behavior has been going on determines how quickly and easily it can be corrected. Chewing correction in older dogs is sometimes more difficult. Destructive chewing usually takes place when the owner is away. One of the staple descriptions we hear in our discussions with clients is the scene they meet upon returning home from work. We have heard horror stories of drapes pulled down, pillows torn open and the contents liberally strewn about, mattresses with their centers hollowed out, plate-glass windows smashed, and wall-to-wall carpeting ripped up. Once the canine mouth gets going, there is no limit to the destruction it can do. Perhaps the most amazing story we have heard concerned a two-year-old Newfoundland–Saint Bernard mix who knocked over the refrigerator, ate five squares of linoleum tile, and ripped the phone out of the wall. When the owners returned, the dog had the audacity to growl at them. Your particular tale of woe probably pales next to this one, so cheer up — something can be done!

We have found a common pattern in owners' reactions after these discoveries. First, they are surprised, even if the dog has done damage before. Next, they are angry and inaugurate a wild chase scene with the dog, sometimes ending with physical punishment but more often with the dog's escape. Even more common, they scold the dog as he cowers in a corner and the owners resign themselves to cleaning up the mess. Many owners explain the behavior by saying, "He hates me for leaving him alone," or a variation on that theme. They are partially right.

The fact is, however, that usually the dog does not have it in for the owner. He may experience frustration on other levels. The factors that produce chewing sometimes originate in the environment, not in the dog. Once these factors are understood, it is easier to solve the problem. The dog's chewing can then be focused on an appropriate object, and the destructive behavior stops.

First, the most sensible approach to destructive behavior is to be realistic. Don't even think of leaving your dog with free rein of the house when you are not present before he is a year old at the earliest. In the meantime, use a crate or other acceptable means of confinement, making sure that your dog gets out for exercise at midday if you cannot be home during the day. Hire someone if necessary. Regular and substantial exercise is a vital component of trustworthy behavior, as is regular practice in the obedience exercises.

Next, trust in the home is earned, never presumed. Once you decide to begin conditioning your dog to more freedom in the house, begin with short departures. A program should be set up whereby the owner leaves for only a few minutes the first day, building up gradually in ensuing sessions. Be sure to temper emotional homecomings or departures, even when they are brief. Greet and say good-bye to your dog quietly, affectionately, but not dramatically. The owner should instill a sense of responsibility in the dog by training him with the words "watch things" or "watch the house" when leaving, and giving gentle praise when arriving home. Each time the dog manages to stay alone for even a short length of time, it is a big victory and a stepping-stone to longer stays.

On the other hand, if on your arrival you catch your dog chewing something inappropriate, correct him, put him on a down-stay, and then focus his attention on an appropriate chew object. We know one owner who installed a reverse peephole in her front door so that after a mock departure, she could look into the living room and see what her Labrador retriever was up to. When the Lab started to get into mischief, the owner simply opened the door and was able to make a perfectly legitimate, well-timed correction. In no time the Lab was conditioned to be entirely trustworthy when alone. Finally, leaving the radio on is another strategy that sometimes helps distract the chewer and make him think his owner is going to return soon (see chapter 23).

Even though some people swear by it, physical discipline for chewing is not effective, unless the dog is corrected in the actual act of chewing. Suggestions of taking the dog to the scene of the crime, focusing her attention on the chewed object (either by picking up the chewed article or bringing the dog's eyes down or up to it), and disciplining the dog under the chin are misguided. Dogs do not register delayed cause and effect, and thus do not understand the meaning of the correction.

A more promising preventative approach blends sensible concrete measures (i.e., chew-proofing the house, using repellents, and set-up situations) with a more fundamental correction of faulty aspects in the dog-owner relationship. If your dog is chewing on something predictable, it is a good idea to attempt to make the object being chewed unpleasant in itself and avoid, if possible, connecting this discipline with the owner. One way to do so is to use repellents; another is to set mousetraps strategically in the chewing area. Don't worry, they won't get stuck on the dog's nose. They can accomplish two objectives: when you are gone, they can act as reminders to the dog to stay away from a given area; when you return home, you can gauge your progress simply by looking to see if the traps went off.

Remember to approach the problem positively, too. Provide the dog with a nylon or rawhide bone. Make this bone a special toy. Play fetch with it. Wiggle it on the ground in an inviting way and let the dog chase it. Keep this bone away from the dog for at least two hours prior to leaving home. Just as you leave, rub the bone between your palms for two minutes to leave your scent on it. Then offer it to the dog as you leave. Make sure the dog sees you offering it. If the dog takes it from you, so much the better. Do not provide the dog with old leather shoes, socks, or personal items. Dogs are not able to tell the difference between old and new, what is permissible to chew and what isn't.

Avoid disciplining chewing on forbidden objects by other techniques, such as taping the chewed object in the dog's mouth (a good way to get bit), hitting the dog with the chewed object (the dog won't make the connection), or tying the chewed object around the dog's neck (the dog might chew on it again), all methods that have been used from time to time.

To correct chewing, first reorder the relationship between you and your dog by establishing yourself as the alpha figure. Obedience-train your dog. Provide effective discipline if you discover a chewing incident. Focus the dog on an acceptable chew toy. All other techniques, such as sprays, mousetraps, screamed or shrieked "nos," and prayers are simply adjuncts to this basic approach.

## Digging

Dogs who are forced out into the backyard, often "for their own good" or "to get fresh air," soon resort to frustration-release activities, and digging is one of the most popular. Since digging is often related to social isolation, the most positive step an owner can take is to allow the animal into the house. Obedience training, diet regulation, and other changes are to no avail unless the dog is also included in the owner's life.

Interviews with owners of diggers often reveal that the dog lives in the backyard "because he drives everyone crazy when he's indoors." Sometimes the owners cannot even remember when their dogs were last allowed inside. In other cases, the dog is allowed inside only at night (when the owner is ready for bed) and excluded from all other activities. Obedience training, emphasizing the down and down-stay, is imperative if the dog is allowed into the house on a steady basis. If you are experiencing a digging problem, begin a program of obedience training immediately. Also, make sure your dog gets generous amounts of vigorous daily exercise. Teach her to play fetch. Include your dog in your activities. Some methods intended to solve digging are filling the holes with water and shoving the dog's face into them, installing chicken wire in the holes, and rigging the holes electrically (very dangerous). We have little confidence in such methods because they usually ignore the underlying problem: the owner's banishing the dog from the social circle.

Digging can, occasionally, be related to breed type. Some Siberian huskies and other sled-dog breeds and mixes seem to enjoy digging cooling holes. Dachshunds and other breeds genetically geared to go underground may show a predilection for digging. Pregnant bitches often begin burrowing activities as they near their whelping date. Most dogs respond well to a program of simple obedience commands and by being included, rather than excluded, from social activities inside the house.

## Jumping Up

Why do dogs jump up on people? Usually they want attention, and since the face is the center of human communication, that is where they first seek that attention. It's an unwanted compliment: "I love you"— canine style. Also, many owners inadvertently encourage this common problem by allowing their new pups to playfully jump up on their legs while the owners reward them with petting. At twelve weeks of age this seems harmless enough, but how the perspective changes six months down the line! An exuberant, seventy-pound "teenager" can easily knock down a child or an elderly owner and assault your guests (however playfully), not to mention ruin your clothing and generally become a nuisance.

The most popular solution is the knee-in-the-chest routine. The trouble with this correction is twofold. First, it should not be the initial correction administered (we use it only as a last resort), and second, it is usually executed so poorly that the point of the correction is completely lost on the dog.

It's far better to start correcting the problem in puppyhood by teaching the sit. Establish a rule for the entire family pack that no jumping is allowed, whether you are wearing dungarees or a white suit. You have to have a consistent policy. Dogs cannot discriminate when it is acceptable to jump and when it is not. When you sense that the pup is ready to jump up, tell her to sit and put your palms out flat in front of her face to block jumping. When the pup sits, praise her warmly. Take the opportunity daily to condition the young pup by having her sit in front of you or other people.

Older dogs may need a more physical approach. Simply grasp her paws when the dog jumps up on you; gently move the paws slightly to the each side and begin moving slowly to keep the dog up on her two hind legs. Don't show any anger. In no time the dog becomes quite uncomfortable and wants to get down. The object is to establish in the dog's mind an unpleasant association with jumping up. Hold the paws a few seconds longer and then finally let the dog down and guide her into a sit. After giving her a firm "stay," jump up and down several times in front of her to give her the opportunity to jump up again, praising her quietly if she holds the sit-stay.

A word of caution about the popular knee-in-the-chest correction. It should be reserved for chronic cases, not for puppies or occasional jumpers, and it should *never* be done by anyone other than the owner(s). To execute it correctly, begin by observing the knee kicks of drum majors at football halftime events. When the dog jumps up on you, immediately grasp its paws and hold them. At the same time bring your knee up into the dog's mid-region and make firm contact. Then say, "No!" and push your arms out to heave the dog away from you. It is not so much the physical force involved that is important as the element of surprise and drama. Again — caution — this is not for puppies or sporadic jumpers. Try the other preventative methods first for them. Domineering owners may seize upon this correction to teach the dog not to jump "once and for all," which may backfire and have a detrimental effect on your overall relationship. Don't try to teach your dog to stop jumping up without teaching her the sit at the same time. Finally, whatever approach you use, be consistent. Remember, one good correction will save you fifty ineffective ones. If you want your dog not to jump, follow through the first time she jumps up.

# Protection Training and Attack Training

If you are considering having your dog attack-trained, ask yourself the following questions:

- Why do you want this kind of training?
- What do you know about this kind of training?
- How many other dogs have you seen with this kind of training?
- Do you understand the difference between attack training, protection training, and Schutzhund training — or do you think that they are all the same?
- Will you or your family be able to control your dog if you have him attack-trained?
- Are you covered by insurance for any harm your attack-trained dog might do?
- Does your dog need this kind of defensive training, or do you?

In our experience, in almost all cases, we have found that owners who seek attack training are misinformed and unqualified to handle such dogs.

Many owners want their dogs attack-trained because they feel threatened. Life in a big city and other crime-infested areas can be frightening, and men and women living alone need protection. But whether formal attack training is in order is another question entirely. Should a dog feel responsible *for* a person or *to* a person? There's a big difference. Dogs can be taught protective techniques that accomplish the owner's goal without going to the extreme of attack training. Even some trainers who specialize in attack training will tell you in candid moments that a good dog defends his owner naturally, without specialized training. Often the mere presence of a dog discourages intruders, and certainly a well-trained dog who can bark on command usually provides all the intimidation an owner may need.

There are several methods of attack training. Most, but not all, consist of a system of heaping "last straws" on the dog, demanding that the dog be put under stress and agitated until he decides "This is it! I'm not going to take it anymore!" resulting in growling, barking, or, eventually, a full-charge attack. This basic agitation is then channeled into verbal and hand signals — key words and gestures — such as "get him," "*fahss*" (in German), and a raised arm, on which is usually worn a protective sleeve. Rarely, if ever, are the stereotypical words *kill* or *attack* used. The dogs are also taught the "out" command — to back off completely. This is a simplification of attack training, and we understand that to many responsible trainers (such as those in law enforcement), it is a sophisticated art.

However, such training presumes that the dog be absolutely sound, since it places him under a considerable amount of stress. As any police officer will tell you, the vast majority of dogs are unsuited for such training — only a small percentage can serve safely and reliably after such training. Consider whether your dog can sustain such stress, and whether you have the requisite knowledge of dogs to handle such training. Most often the answer will be a blunt no. More seriously, for such unsuited dogs, attack training most likely will have very negative side effects that could eventually endanger their lives. The contemporary legal system has little patience with vicious dogs. Also, though some trainers may be honest with you and tell you if your dog cannot take this kind of training or if they feel you could not handle a dog with this kind of training, others may not. Some trainers "attack-train" almost any dog, given enough time, the right method, and a paying owner, so don't be naive.

For example, protection training often backfires. It does not create a robot who will be automatically responsive to all commands. In fact, because owners who opt for such training often are interested only in the protective aspects, they can easily disregard the essential foundation of absolute obedience and the basic soundness of the dog upon which it must be built. The relationship itself is also of secondary value. Such imbalance can easily lead to "learned aggression" (see chapter 43, on aggression), which many owners have difficulty controlling. Once the genie is out of the bottle, accidents often result —

as headlines in today's newspapers sadly attest. Retraining in such instances is difficult, if not impossible.

An attack-trained dog is like a loaded gun. It should be handled only by experienced people in appropriate situations. We commend the use of German shepherd dogs and other breeds in police units across the country. When used defensively, they can be of great value. These dogs belong to qualified handlers who live with them twenty-four hours a day. They are trained to high standards and are sound genetically. Laymen, on the other hand, should not have attack-trained dogs. They are not qualified to handle the animals.

Although we do not recommend protection or attack training, Schutzhund training, first developed in Europe, might hold out a possibility for viable defense training within the scope of a nonprofessional, but even here serious commitment on the part of the handler is essential. It is a threefold obedience system, encompassing tracking, obedience, and protection skills. In true Schutzhund training, no one area is overplayed or allowed to get out of balance with the others. Significantly, this training system treats the dog as a whole. It proceeds from the correct premise that the dog is responsible to the owner. Schutzhund handlers make the protection phase of training more of a game than anything else, even though it can be of use in real-life situations. Although there are eccentrics within any movement, and currently much division within this particular one, Schutzhund training is a sound method and may be of value to your dog.

However, you must begin with a sound, healthy, discriminating animal, usually from one of the working breeds. You must delve deeply into the training method to become a qualified handler. This will mean joining a Schutzhund club.* It is not possible to explain Schutzhund training in full here, but it is a good possibility for those who feel they need protection training.

To sum up, you will notice that many ads for attack-trained dogs play on owners' basic insecurities, promising "freedom from worry" and "complete safety." Methods of training that do not relate to the whole dog and that fail to educate the owner properly do a disservice to dogs and society. If you are interested in protection training, please think twice.

*For information on the Schutzhund movement in the United States, contact United Schutzhund Clubs of America, 3810 Paule Ave., St. Louis, MO 63125-1718.

# 42

# Alarm Barking

In the preceding chapter we come down hard on the standard types of attack training. But we are not against pets announcing the presence of visitors and, if necessary, protecting their owners from physical harm. We are breeders of German shepherd dogs and work hard to preserve the structural and temperamental integrity of the breed. We value that our dogs have the instinctual drive to bark and defend. That's usually enough to ward off intruders.

Owners who want their pets to protect them should teach their dogs to speak and then learn to use them psychologically to the best advantage. The major factor in many a tight squeeze, such as a robbery or assault attempt, is the victim's ability to bluff the attacker. Bluffing with your dog means portraying the dog as a true friend and defender, and not as the weak, timid critter he may actually be.

If you are being followed or bothered, it's a good ploy to bend over dramatically and whisper something to the dog. For a direct approach, hold the dog tightly by the training collar and say, "Not yet, not yet," to bluff intruders. The average intruder has no way of reading a dog correctly and cannot tell that you are bluffing.

If you want your dog to bark on command, begin teaching him as a puppy. Hold a treat over the dog's head and when he sits, tell him to "speak!" Encourage the dog to vocalize, and if he gives you even a whine or whimper, treat him promptly. Some young puppies fall naturally into the game, and older dogs sometimes take more time to train.

When you are confronted with an unpleasant situation (assuming that your dog is not already barking because of it), you can fake holding a treat and get a few barks that way. The other person usually cannot decipher whether it is a friendly or unfriendly bark, and most likely will not want to take the chance of finding out.

To train your dog to bark aggressively at an intruder, set up situations in which someone unknown to the dog can help you. Have your accomplice cover his or her face with a nylon and wear a heavy coat and strange hat. The hat is the main thing that triggers the barking reflex in many dogs — make sure it is a bizarre hat. Have this decoy make some motion outside the house to set the dog off — rattling a doorknob, jimmying a lock. You should be sitting quietly, reading, with the dog nearby, preferably lying near you. When you hear the sound, alert and look at your dog. If your dog gives a slight grump, consider the exercise over for that day. If the dog does not react, go over to the door with the dog, encouraging him to investigate. As the dog sees the stranger, he may growl or bark. Have the stranger reel backward, linger for a moment, and then run away, as if running away from the ferocious dog. Then kneel and praise the dog for "saving" you from the intruder. Repeat this procedure on a weekly basis, but don't let things go too far. All you want is a warning growl or bark. After such setups it is a good idea to balance them with introductions to friends who come to the house. This way, your dog's protectiveness becomes something entirely within your control. If you already have problems with these two activities, or if your dog has a problem with aggressive behavior, don't condition him further with these setups. Remember, too, that many breeds have a built-in propensity for protective behavior; it is a matter of waiting for this trait to mature. Don't expect a three-month-old puppy to be Rin-Tin-Tin.

# 43

# Aggressive Behavior, or How to Deal with a Canine Terrorist

Canine aggression is one of the most frequent problems we deal with when dog owners bring their dogs to us. After seeing hundreds of aggressive dogs, we have come to certain conclusions about the nature of such aggression. First, it is natural canine behavior to establish and preserve rank within a pack, for a dog to protect himself from anything he perceives as threatening, to procure and protect food, to defend territory, and to establish dominance in a variety of situations. Genetic history and selective breeding can increase or decrease the tendency toward aggression in breeds and individual dogs, as can environmental circumstances. To the untrained eye, aggressive reactions are often seemingly unprovoked, unpredictable, and unexplainable, but when examined in the overall context of canine behavior, they become more understandable, if no less serious.

It is important not to underestimate the scope of canine aggression. It has been, and continues to be, a serious problem that is present in all canine environments. In urban centers, dogs who roam freely resist anyone attempting to round them up; they wander about, picking up new "recruits" along the way, and in the course of packlike dynamics the more dominant and aggressive initiate the others into increasingly aggressive behavior. In the country, packs of dogs loot chicken houses, attack sheep, run livestock, and challenge farmers who attempt to catch them. And every suburban neighborhood probably has at least one or more well-known "mean dogs" whose aggressive behavior is fabled by residents. Let's not kid ourselves: facing an aggressive dog is frightening, even for a professional. Ask the question "Have you ever been bitten by a dog?" of practically anyone, and the answer, if yes, will also include vivid details of the incident. Dog bites are not easily forgotten. Often the context is predictable: a situation involving a strange dog who was unsupervised at the time, along with some action by the bitten party that the dog interpreted as threatening.

Most owners have no desire to live with an aggressive dog, and certainly not with one they cannot control. The sad truth is that many owners put up with such circumstances because they did not act swiftly and decisively when the problem first manifested itself, usually in the first six or seven months of the dog's life. Procrastination and denial are major contributors to a mature expression of aggressive behavior. Aggression rarely becomes a serious problem overnight: it develops through incidents of escalating intensity that are not checked by the owner. Throughout this book we have emphasized the role of prevention in raising a safe and healthy companion. It is much easier dealing with budding forms of aggression in a puppy than full-blown aggression in an adult. Following a dedicated program of socialization, training, and play, as well as setting clear and consistent limits to inevitable displays of juvenile aggression, is the surest path to preventing aggression from becoming a real problem in the adult dog.

That said, what of the older dog manifesting one or more forms of aggressive behavior? Is there no hope? Training techniques alone cannot solve every case of aggression. But good training techniques *and* counseling can diagnose and evaluate aggressive behavior and attempt counter-conditioning. We cannot stress enough that if you are having a problem with aggressive behavior, see a trained professional as soon as possible. Preferably, see an experienced trainer or behaviorist on a one-to-one basis so that he or she can work individually with your dog. Do not attempt to "cure" aggressive behavior simply by enrolling in a park obedience course. Go to an expert professional who can evaluate your dog's behavior individually, preferably one who can work with you in a hands-on way. Get such help as soon as possible. Don't wait, thinking that the behavior will go away or get better. It rarely does without training. The behavior usually gets worse.

It is important that the trainer/behaviorist know all the details of any aggressive incidents. It is essential that you evaluate the incidents, preferably with the professional's help. It is possible to work alone, but it is more difficult. Get a notebook and write as objective and un-emotional a description as possible of what happened in each incident. Describe the people involved, the time of day, the actual split-second sequence of events (as best you can recall), and what discipline (if any) you administered. Note where the incident took place and the extent of the damage. It is important to know whether the dog broke the victim's skin *(were stitches required?)*, or nipped, scratched, or inflicted the damage with his mouth or paws. Many times we've had clients report mouth bites only to discover later, when taking a case history, that the injury came from the dog's toenails or paws, not from the mouth at all. Check and recheck. Don't justify or condemn the dog's actions. Simply record the facts. This is the first step in deter-mining the seriousness of the problem.

After you figure out the sequence of exactly what happened each time, it is now possible to attempt to categorize the aggressive be-havior. Contemporary animal behaviorists such as Benjamin Hart, William Campbell, and Nicholas Dodman confirm what experienced trainers have learned from their day-to-day work with people and their dogs: there are different types of canine aggression.* Listing the most noteworthy may help you determine precisely what kind of behavior your dog exhibits. Keep in mind that your dog may be ex-pressing more than one type of aggression.

## Fear or Shyness Aggression

By far the most common type, fear aggression occurs equally in males and females and is usually seen in situations of stress and noise, when the animal would prefer to escape (leave the area) if at all possible. Instead, he is forced into a confrontation with another dog or a human. For instance, an owner who drags a reluctant Fido up to "meet" strangers, instead of conditioning the dog more slowly to accept company, invites an incident of fear aggression. Most incidents happen in stress situations when a dog's initial display of fear is pushed into actual aggression. Think of the conditioning process as

*See the Select Reading List.

like getting into a hot bath: wise are they who go slow. Obedience training — at least to the come, sit, and stay level — is necessary, since the lack of leadership often contributes to aggressive behavior.

The fearful dog needs to have confidence in his owner before he can even begin to negotiate meeting strangers. Then, instead of letting the dog meet the stranger head-on, we first work in the general vicinity, moving closer as the dog gets used to things. We may sit the dog several feet from the stranger, praising the dog as he complies. Several sessions may be needed to work positively toward an actual encounter. Be patient.

As far as dietary changes that may be in order, behaviorists and trainers are divided over a high- or low-protein diet, and it seems that more research is needed to come to more definitive conclusions. But while the relationship of high- or low-protein diet to behavior change is uncertain, everyone agrees that obedience training is an absolute part of any solution to aggressive behavior. Don't think you can solve aggression independent of a consistent regimen of obedience training, and don't use physical discipline to correct such behavior. It only compounds the fear.

In terms of desensitizing fear aggression itself, the round-robin recall is often very helpful once the dog can tolerate it. Some veterinarians have experimented with drug therapy for this kind of aggression and have had varying degrees of success. Currently Inderal, Prozac, and BuSpar are the drugs of choice. However, do not attempt to administer drugs to your dog without veterinary/behavioral assistance. Drug therapy can sometimes backfire or simply mask underlying problems in the dog-owner relationship that are the root cause of the behavior.

## Territorial Aggression

The second most common form of aggression is the front-yard and backyard variety. To a degree, in-house and in-car aggression is natural. Owners often put their dogs in a bind, encouraging territorial aggression but discouraging it when they find it excessive. The dog may be confused. Alarm barking does not mean the dog should bark at everyone. You must be in control of his barking and have the ability to turn it off at will with a sign-off phrase like "no more!" or "okay, that's enough." Discourage any kind of fence-running out in the yard, if necessary using a bark collar to stop such tendencies.

From the start, the best rule to follow is to inhibit territorial aggression in pups with a stern "no" and friendly exposure to strangers. This type of response usually crops up in the pups about the fourth or fifth month. Be on the lookout for such a reaction and clip it immediately. Do not encourage territorial barking; take advantage of introducing your pup to mail carriers, meter readers, and other regular "intruders" early on, setting up positive encounters that establish a context of familiarity for your pup. For example, in addition to setting up friendly introductions in which both owner and outsider act happy around the dog, invite such outsiders to play fetch with the puppy or dog, neutralizing his uptight response. Also, pay attention to your own body language: consciously greet visitors in a happy way. Obedience-train the dog for control.

For older dogs with an already existing problem, begin a program of desensitization. Set up situations that progress from a measured response to the doorbell's ring (i.e., a brief bark that stops with the owner's command), to opening the door, to an actual introduction, first with someone the dog knows and loves, later with a stranger. Move from one stage to the next only when the dog is controlled and responsive in the existing stage. Only when your relationship with your dog is confident, with little fear that your dog may try to bite you, should you discipline aggressive displays according to the principles described in chapter 10. After disciplining, be sure to go immediately into a sequence of obedience commands that reassert your position of leadership, then try the setup once again. Often your dog is much more compliant this time around. If you suspect that your dog may try to bite you if you discipline him, seek private help with a professional. Most likely you will need to go with a nonconfrontational, "nothing in life is free" approach to your dog that works one step at a time and demands patience and consistency to be successful. Your dog will be expected to earn everything it receives: praise, attention, affection. . . . The professional/behaviorist will outline this for you in exacting detail.

## Intermale Aggression

Males generally fight other males. The problem is an inescapable canine situation related to testosterone secretion and the environment in which the males are raised. For example, in a breeding program such as ours, males must be kept separated because they view one another as competitors. However, this complaint is also frequent among dog owners whose dogs meet up with free-roaming neighborhood males. To control this, defecation and urination by a male must be restricted to the dog's immediate area before you go on a walk. This is so that he does not mark off the whole neighborhood as his private domain, which easily leads to confrontations with other males he interprets as trespassing on his kingdom. Neutering helps reduce this behavioral trait, and to make your dog less dominant and threatening in the eyes of other males. Finally, take sensible precautions. If you know your dog to be dog-aggressive, use a muzzle when you take a walk.

## Aggressive Response to Teasing

This is a situation when children, and occasionally adults, taunt a dog until he retaliates. It is sometimes common when neighborhood gangs team up to pick on a particular dog, often when the owners are away, but can also occur when an individual child relieves boredom by teasing the family pet. In addition to supervising interactions between children and dogs, it is of paramount importance that parents educate their children in appropriate etiquette: children should be cautioned not to scream around dogs. They should never chase a dog, even one they know, nor should they ever pet a strange dog. Adults who tease dogs or accept "dares" to approach a dog get what they ask for. One young man accepted a dare to stick his hand into a car occupied by a Doberman he did not know. He was bitten and wanted to know why!

## Pain Aggression

Pain aggression often occurs at a vet's office when the dog is given a shot, wheels around, and bites the vet or vet assistant. Being elevated onto an examination table is often enough to set some dogs off into a bite reaction when injected. Though veterinarians may insist on examination-table treatment, many such isolated bites and aggressive displays can be solved by on-the-floor exams. If you have a problem with this type of aggression, ask your vet in advance to treat your dog on the floor if elevation to the exam table is not essential. Only a veterinarian with a slipped disc should refuse.

Another type of aggression, which we have frequently seen, is the result of pain induced by hip dysplasia, auto accidents, or other injuries. For animals with these conditions in their background, aggression when getting up to a standing position, especially if forced, is a possibility. We have seen many cases in which large breeds have been involved in biting incidents when a child sits on the dog's hindquarters when playing, causing the dysplastic or previously injured dog to wheel around in pain and bite. Prevention is key here: respect the individual dog's condition and don't expect him to tolerate pain brought on by needless roughhousing or surprise contact. The only way to check for dysplasia or injury is to go to a vet and have your animal x-rayed. Treatment can often be as simple as aspirin administered twice daily, or as complicated as an operation to relieve the tension.

## Learned or "Trained" Aggression

Chapter 41 in this book deals with the pitfalls of protection and attack training, not the least of which is the danger that the dog might misuse his acquired aggressive skills. Owners considering attack or protection training certainly should think twice about it. "Deprogramming," while possible, is difficult. Rehabilitation of these canine soldiers usually entails a prolonged separation from the owner; an emphasis on animated, happy training sessions; and avoidance of the cue words and hand signals the dog might have been conditioned to respond to with aggressive behavior.

## Genetic Aggression

As professional breeders, we have had experience with the possibility of genetic aggression that some trainers do not have. This is an extremely difficult area to diagnose. For instance, genetic aggression often looks like fear or shyness aggression. Good trainers take into account the possibility that the aggressive behavior has genetic roots. They take the time to investigate the dog's bloodlines, checking the pedigree with the owner (and, if possible, the breeder) as well as the owner's recollections of the behavior of the sire and dam. "Certified" pedigrees with no degree or title information on the ancestors should be viewed suspiciously. Aside from puppy-mill and backyard breedings, almost every breed has what breeders call "freak bloodlines" that produce dogs with a propensity for aggressive reactions. Particularly when the aggression appears in the young dog, chances are good that one or both parents had trouble in this area. Experience can sensitize a trainer to detect genetically based aggression, but it is still very important for you to provide the trainer with all possible information about the dog's ancestors if you suspect this possibility.

If genetic aggression proves to be the diagnosis, there is a serious and difficult decision to be made. If we are dealing with a situation in which the dog's aggression appears to be dangerous and unpredictable, and the dog has badly bitten a number of times, euthanasia has to be considered. Any sort of remedial training will at best only mask a deeper problem that most likely will erupt spontaneously in manner that could injure someone seriously. As we shall see in chapter 46, there are times when the most loving and humane action an owner can take is to end a dog's life peacefully.

Maternal aggression, which crops up when a bitch has a litter and overdefends it from all comers, seems to be passed from mother to daughter. If this was a problem with the mother of your female and you plan to breed the daughter, reconsider your plans.

Again, we stress that if you are interested in a puppy, buy only from a reputable breeder. Try to meet the sire and dam of your puppy personally — not from behind a kennel cage, but in normal house circumstances if at all possible. Many dogs bark and look aggressive when "behind bars." If the breeder is hesitant to take the dog out of the kennel, or if the sire or dam shows an aggressive response, reconsider buying a puppy from their litter. Don't think that an aggressive father instills courage in your puppy. You may get much more than you bargained for. Purchase a sound puppy from sound parents and a conscientious breeder. Obtain a guarantee of temperament. This is the best way to sidestep the chance of genetic aggression.

# 44

# Behavior In and Out of Cars

## Riding in Cars

We often hear about the dog who "can't ride in a car." Dogs who vomit, whirl around, bark incessantly, or attempt to jump out of a moving vehicle are becoming a more frequent training problem. As urban and suburban auto use has risen, we find more dogs who are unable to adjust to the stress of riding in a car. Usually these dogs were not properly conditioned to riding as young pups. The poorly trained puppy develops a fear of cars at worst, and at best a lifeless resignation to riding. If we wish to include our dogs in our lives in an optimal way, we'll need to accustom them to riding in cars.

### The First Ride

Ask the breeder whether the pup has been exposed to riding in a car. Tell the breeder that you prefer the pup's first ride not to be the ride home with you. Some breeders use foresight and load up a whole litter early after weaning to take them out for a short spin several times before sending them to their new homes. On adoption day, make sure the puppy has not eaten for at least three hours prior to leaving with you. If you have children, explain to them that the puppy will need quiet on the way home. During the pup's ride home with you, avoid any unnecessary coddling. Place the pup on the floor or on the seat next to you on top of a thick pile of spread-out newspapers. Do not respond to whining by petting (rewarding) the puppy. You cannot punish the puppy for any vomiting that may occur now — it is an involuntary reaction. If you scold and discipline the pup, you make him even more nervous about riding in cars. We often suggest that our clients simply drive to the bottom of our hill, stop under a shade tree, and rest a moment with the pup before continuing home. If you are going more than twenty miles, stop again for five minutes to give the pup's stomach a rest. The point is to try to get the pup home before he gets sick.

## Teaching the Dog to Ride

Once you are home, begin acclimating the pup to riding in the car. For your initial trips, choose a smooth, straight road. Withhold food and water for at least three hours beforehand. Prior to starting out, let the pup sit in the parked car for five minutes. Proceed in the manner described above with short trips, ending in a play session. Gradually lengthen your trips until your dog can go farther than twenty miles without any sign of heavy salivating or indication of vomiting.

Whether puppy or older dog, your attitude about riding in the car affects your dog. Owners who coddle their dogs, overuse tranquilizers, or sympathize with stress-whiners actually encourage car problems. You should praise the dog excitedly as you near the car. As the dog jumps in (or as you load it in), give the dog a good deal of physical and verbal praise. After that, your job is to drive and the dog's job is to ride.

For safety's sake, use either a crate, canine safety harnesses, or a vehicle barrier to protect your dog against sudden stops or accidents. Crates fit conveniently in the back of station wagons and SUVs, and sometimes can fit on the backseat of a smaller car. Safety harnesses allow the dog to sit comfortably and securely on the front or back seat of any vehicle, and the vehicle barrier fits in vans, SUVs, and station wagons. There are simply too many possibilities for the unprotected dog to get injured with a quick stop to risk leaving your dog free in the car. Do not tolerate any barking from young puppies. With a friend driving, either use a shaker can (twenty pennies in a soda can — dogs hate the sound) or clench the pup's mouth shut and give a slight shake with a "nah."

## Older Dogs with Car Problems

If you have an older dog with an established barking problem, try this method. Begin on a normal trip with the dog confined or restrained.

Deliberately drive past situations that you know will trigger a barking reaction from your dog. Just as the dog begins to bark, say, "No!" loudly and use a blast of an aerosol air horn. Doing so is safe and allows you to keep your eyes on the road. The sound is so disconcerting to the dog that it will break the dog's focus and quiet the dog.

Now drive on about a block and turn around and drive right past the same distraction as before. "Read" your dog's reaction in the rear-view mirror. If the dog appears eager to bark, slow down, and warn the dog. At the first sign of any whining or barking, repeat the routine.

You may wonder whether this procedure would be more effective if you drove and had a partner execute the corrections. It might be, but we have seen many cases in which a recalcitrant car barker clams up when the driver is accompanied, only to burst out in a flurry of barking when taken on a "solo flight" with one person who is busy driving.

One last piece of advice: if your dog is having a hysterical barking fit in the car, refrain from any yelling or screaming at your dog. Not only will the dog not hear it (it is too focused on the distraction), it will increase the level of your own stress. Condition your pup to driving in the car at a young age, and use the air horn for more ingrained problems.

## Car Chasing, Barking at Cars

Dogs who chase cars usually suffer from hyperactivity or (more probably) sheer boredom. They are instinctively oriented to chasing down things that move, which in the wild was how they obtained their food. Additionally, dogs can be protecting their own territory, driving off the intruder with an explosion of barking. Such behavior is reinforced each time they chase a car and it "runs away." For a dog, such "successes" swell her pride and lead to repeated incidents of car or bicycle chasing that can be extremely dangerous. Even after a dog has suffered a serious injury, chasing behavior can persist, risking further injury or death.

The simplest solution is to prevent your dog from running loose on your property unsupervised (something we never recommend if the property is not fenced-in). Think preventatively: have the dog trail a long leash or rope from her collar when you are playing fetch if you suspect an inclination toward chasing and have not yet had the chance to address the problem directly. Then you can grab the rope if you have to stop the dog. But understand that the problem can express itself on simple walks along a road even with the dog on leash; suddenly she may bolt out toward the moving vehicle or bicycle. Should this happen, quickly move in the opposite direction, letting the dog receive the dramatic correction when she hits the end of the leash.

However, the best policy is to clip this behavior quickly when you first notice it in your puppy. Often puppies simply bark at cars or other moving objects, such as bicycles. The pup perceives the moving object as threatening. Pups should be exposed to moving vehicles in a controlled way, on leash, at a young age. Walk along a busy street with the pup at your side. Stop and encourage the pup if she looks hesitant about cars. If she shows too much interest, move away in a mock-frightened way. Never allow the pup to be called by persons in a car or on a bicycle.

Chronic car, bicycle, horse, or other-dog chasers need a correctional regimen that spans several areas. Popular corrections like heaving a water balloon at the dog from a car window, throwing BBs in his face, or using a squirt gun to ward off car attacks may be effective with some dogs, but they can too often be simply hit-or-miss measures. We don't place much confidence in them. A better strategy centers on the owner in a twofold approach. First, work on a renewed relationship with the dog, emphasizing the obedience exercises and the leadership of the dog owner. The dog must be trained to the sit, stay, and come level, at least. Preferably, the dog also should learn to heel and lie down on command. Second, set up a staged confrontation between dog and owner when the dog is chasing a car.

To concoct this confrontation, leave your pooch with a leash attached for two days. Just let the dog walk around with the leash dragging behind. Begin by having daily obedience sessions, emphasizing sit, stay, and come. Leave the leash on after the session. You will need it later. Secure the help of a neighbor or friend who is willing to drive by your property for a setup. It is best if this helper has a car that is unfamiliar to the dog. Explain your dog's behavior to your helper before this training exercise. Map out the dog's behavior on a sheet of paper, noting the position of the car and dog when the chasing normally begins. Some dogs are erratic. They may begin to chase a car when it rounds the bend a block from the house, but other times act excited only when a car is directly in front of the house. Still others behave like clockwork, charging when each car is lined up in a familiar position. Whatever your dog's habits, prepare your helper for them so that he or she will know when to brake.

When you are ready for your confrontation, have your dog out in that part of the yard where the chasing usually originates, trailing a long leash or rope. To prevent any kind of rope burn, wear gloves. Allow your dog the opportunity to walk freely trailing the rope, but stay close enough to him that you can pick up the rope quickly. Now have your friend drive the car slowly by the border of your property. When your dog begins to chase after the car, pick up the end of the rope and immediately begin running in the opposite direction (away from the car), holding the rope tight. When your dog hits the end of the line, he may well go flying in the air in surprise. Tap your leg and call him to you. Do not discipline him physically when he arrives. Simply put him on a down-stay and have the driver repeat the temptation several times, alternating starting positions with the dog. Since your dog will not be allowed to run freely in the yard without supervision, your monitoring the situation will teach him quickly to ignore passing cars.

Bear in mind that car, bicycle, or people chasing is often the result of two canine frustrations. First, the bored dog who is ostracized from the house for any considerable length of time eventually vents his frustration in some way — chewing, digging, or chasing being the most common. Second, the dog hooked up to a chain or kept behind a fence affording a full view of traffic often develops what is called barrier frustration. This pent-up tension is then aimed at any and all free-moving objects. Of course, a combination of boredom and barrier frustration is twice as mind-boggling for the dog. Part of the solution to chasing lies in somehow eliminating these conditions. The simplest way is to bring the dog into the house, where he belongs. If no one is home during the day, explore the possibility of enclosing a yard and installing a dog door. Some dogs may not need access to the outside at all and can be kept inside except for supervised exercise and defecation periods. The biggest holdup preventing these simple solutions is the owner who believes the myth that dogs "need to be outdoors" or that "dogs need exercise all day and should be allowed to run free." Believers of these myths might as well resign themselves to the possibility of a lifetime of coping with chronic chasing activities . . . or a dead dog.

# 45

# Social Implications of Training

The Humane Society of the United States estimates that there are 150 million owned dogs and cats in the country. Though the society has done remarkable work in education and pet adoption and has seen a 30 to 60 percent decline in euthanasia in recent years, last year alone the nation's shelters still had to euthanize 4 to 6 million dogs and cats. The pet population in this country has been and continues to be out of control.

It is the responsibility of all dog owners to control the reproductive potential of their dogs, whether male or female. Spaying or neutering your dog is imperative unless you have serious breeding plans. If you do, please consider them carefully. Litters are hard work — take it from us; we've whelped and raised many over the years. The puppies demand time and patience. They must be socialized, trained, and placed in proper homes. It is serious business. By no means breed because you feel doing so would be good for the dog, or because you want your children to see the miracle of birth.

When you train your dog, you are helping to stem the pet population explosion. Dogs that are obedience-trained have a healthy rapport with their owners. These dogs are not allowed to roam free and are managed responsibly, which is one important way of preventing unwanted backyard breeding. Properly trained dogs serve an important public function, since they help to balance out, in the public's eye, the large number of badly behaved dogs. Obedience training encourages good breeding. Poor genetic specimens usually do not train well, but dogs who are bred for brains and beauty do. They act as advertisements for responsible breeders and divert business from puppy mills and pet shops. Obedience training is an ethical imperative that not only improves the lot of dog and owner alike but can also help us control the unwanted pet population.

# 46

# When a Dog Dies:
# Facing the Death of a Dog

In the voluminous literature on canine health and training, it is still the case that the death of a dog and how the dog owner reacts are subjects often avoided. Yet the death of a beloved pet is a reality dog owners may have to face more than once. Part of training involves training yourself to accept the inevitability of your pet's death.

In dog owner consultations we have had an opportunity to discuss with many clients the death or imminent death of their pets, whether by natural causes, accidents, or euthanasia. It is never an easy experience for client or counselor. We feel that an owner needs someone to talk to and should feel that the person to whom he or she is talking cares about the dog and the fact that the dog is dead, dying, or needs to be euthanized. Although the last thing someone needs is a cold, clinical approach, sentimentality should not be wheeled in, either.

It is a well-known phenomenon that some professionals who work constantly in the face of death (human or animal) can develop an unconscious callousness and insensitivity to death itself and possibly to those closest to the dead and dying. This is a common way of coping with the constant onslaught of tragedy by veterinarians and others dedicated to protecting life. To care is to embrace one's own vulnerability along with a lasting sensitivity to the needs of others that comes with experience and reflection.

Most pet deaths are quick and painless, and the pet owner does not really have time to absorb what has happened. Shocked by this quickness, they are sometimes guilt-ridden, angry, depressed. They may be suspicious of the veterinary profession or may blame the breeder for producing a poor dog, or the trainer for failing to modify behavior that eventually leads to death. For instance, the dog of one former client ran out in front of a car after taking a training course. The client told the trainer, "If you had taught my dog to come when called, he wouldn't have been hit by that car. You killed him." These are emotional reactions that must be treated with understanding and compassion.

Other owners swear never to have a pet again. Since we live in a society that teaches its members to become emotionally involved with their pets to an intense and quite natural degree, reactions of disbelief, anger, and depression are common. In reality, we know that we usually outlive our pets — unless we are very old ourselves. Those are the facts of life. Dogs simply do not live as long as humans. Yet we still must work through the feelings of loss.

Perhaps the saddest deaths are those of puppies, and next, the death of old companion animals who have lived with their masters for many years. All of us perceive puppies as fragile, mischievous, innocent, and delightful, yet even when we try to take proper care to protect a pup, accidents and disease will take their toll. Owners sometimes blame themselves needlessly for the poor supervision, improper nutrition, or other causes they perceive as the reason for their pup's death. If death results from internal causes or shortly after purchasing the pup, they may blame the breeder for selling a "defective" puppy — even if the breeder had no way of controlling or even knowing about the condition in advance. Though it is fortunately possible to start with a new puppy, the biggest tragedy is the short-circuiting of the young life and the unresolved question of the pet's potential. As with a child who dies prematurely, the owner will always wonder, "What kind of dog would my pup have been?"

When an older dog dies, it might be assumed that the owner can handle the death with equanimity. This is not always so. If euthanasia is involved, the decision can be very difficult. It might be a course of action the owner never expected to have to take. He or she might have expected the pet's natural demise. However, it is important for the owner to understand that true compassion may require such a decision. We were once asked by a television director whether we ever had to euthanize a dog. When we explained that we had, he asked, "What's the difference between putting a dog down and a human being?" The answer lies in the nature of each. Even in the most extreme circumstances of suffering, disability, or disease, a human being can still grow, can still exercise reason and become "more human." With a dog the situation is different. Not being a rational creature, the dog has no capacity to understand his suffering. He simply suffers. When such suffering prevents him from living "as a dog," from moving and doing the things dogs do, the most humane and compassionate thing to do is to put the dog to sleep. It is never right to keep such a dog alive simply for our own emotional needs. Understanding this, however, does not make the decision any easier. It is difficult enough when the animal dies of natural causes — the emptiness and gap in a household can be devastating. The monks at New Skete started their breeding program with their desire to purchase another pup after their original German shepherd, Kyr, was gone. The house was too empty without a dog.

## Getting Another Dog

When a pet dies, the immediate reaction of many owners is to run out and get another. We have had people arrive at our breeding kennels in tears, on the rebound from the pound or veterinarian's office. They explain that they just lost their dog and want one "just like him." We have found that it is usually wiser for these owners to wait before getting another pet. The sad owner projects onto the new dog all of the qualities and talents of the deceased pet, forgetting that each animal is an individual. Occasionally, a new dog develops behavioral problems and does not work out in the new situation. The owner may have unrealistic expectations of the dog or may constantly compare it with the former pet, and the resulting owner-dog relationship is off to a bad start.

Although it is good to plan to get another dog at some point, we've found it better to wait awhile before actually bringing one into the household. In the interim, memories of the old pet start to fade and longing for a new one increases. Children, especially, begin to agitate for a new pet. The owners may have an opportunity to talk about the old pet with a veterinarian or some other sympathetic parties. The veterinarian may be able to explain fully the medical causes of the pet's death, and give advice on what to avoid with a new one. Meanwhile, the scent of the old animal fades, which makes it easier on the new dog, since he could wonder why he smells, but does not see, the other dog.

## The Decision to Euthanize

The decision to euthanize must be the owner's, but almost always there are others working on an advisory level who may be involved. These people — veterinarians, trainers, specialists, animal-shelter personnel, and friends — must be especially sensitive to what the owner faces before and after euthanasia. One cardinal rule is that the owner must make the decision, not the advisers. The advisers' role is one of honesty — to explain alternatives.

If euthanasia is indicated and chosen from these alternatives, it is the veterinarian's or specialist's role to expedite matters as quickly and humanely as possible. Although simple good-byes are needed and wanted by some, not all owners want to say good-bye to their pets. This is especially true if the dog is being euthanized for behavioral reasons, particularly if aggression is a factor. Advisers should not force an emotional good-bye scene. Those acting in an advisory or helping capacity should make the death as painless as possible. If the injection method is used, it is good to explain this tactfully to the owners and emphasize the lack of pain or suffering, the idea that the pet simply falls asleep. This helps the owner to know the truth and to get through the subsequent emotions.

Obviously, euthanasia is not easy for anyone, including the veterinarian or animal worker who has to perform it. Some veterinarians will not euthanize pets simply because the owners request it, especially if the owner says that the dog has behavioral problems but the owner has not sought out training. They may refer the owner to a competent trainer. Veterinarians are dedicated to preserving animal life, not destroying it. So, if you have not worked with a particular veterinarian before, and your first contact is to request that your dog be put to sleep, you may get a flat no.

We encourage pet owners to be realistic about euthanasia. It can be a difficult decision, but time heals the wound and life moves on. It is important to focus on your responsibility to your dog as a friend, to be open to making a decision for him that he cannot make for himself and to be grateful for the time you spent together. Be clear: the humane decision you make as an owner of a pet who is suffering or has serious behavioral problems, while difficult, is entirely different than the staggering legacy of irresponsibility represented by the cavalier extermination of millions of potentially sound companion animals by owners who simply have grown bored with them, who have no time to care for them. In our country, millions of unwanted cats and dogs fall into this category. These throngs of unwanted animals, with no possibilities of homes or owners, represent an abominable waste of life. They are a shocking indication of our lack of reverence for life.

The idea of pet burial is an individual question. For those interested, there are pet cemeteries and the simple rituals of burial. Whether your pet dies at home or not, most veterinary hospitals maintain crematoria where the remains can be decently disposed of. For most people, photographs in an album or a framed portrait help deal with the loss in a healthy way. There are many who prefer to keep the memory of their pet alive in their own minds, privately and discreetly. Others, especially children, benefit from sharing their memories and sense of loss by talking with their family and contributing to a scrapbook of memories with drawings, photos, poems, and essays.

# A Parting Word

# 47

# Dogs and the New Consciousness

*What is mankind without animals? If animals were to vanish from the face of the earth, mankind would perish in the solitude of his spirit. Whatever happens to animals, happens also to mankind. All things are interwoven; whatever touches the earth, touches mankind also.*
— ANONYMOUS, 1855

*Look, I am doing something new, now it emerges; can you not see it? Yes, I am making a road in the desert and rivers in the wastelands. The wild animals will honor me, the jackals and the ostriches, for bestowing water in the desert and rivers in the wastelands for my people.*
— ISAIAH 43:19, 20

Throughout this book we have stressed that a dog is a social being genetically geared to respond in submissive as well as dominant ways to human beings and other dogs. We have emphasized that as his caretaker, you should conceive of yourself as your dog's guide and alpha figure, including him in your activities as much as possible. Once you grasp the importance of these ideas, you should be able to enjoy building a healthy relationship with your pet.

However, there is much more to be gained through your relationship with your dog, if you but open yourself to the possibilities. Your dog can provide you with a unique access to the natural world, helping you to expand your capacity for aesthetic appreciation, warmth, and enjoyment, thus rooting you in deeper realities. In a world grown increasingly artificial and plastic, we are dangerously out of touch with the natural environment that sustains us, and the effect of this detachment has been to create a wasteland of spiritual aridity and alienation. Most people do not suspect that their relationship with their dog can provide a connection to a deeper, more integrated view of the universe.

Our experience teaches us that relating to a dog can be profoundly spiritual. In this book we have avoided religion and religious jargon, and we generally avoid mixing religion with dog training. Nevertheless, many of the ideas we hold about dogs have a philosophical and spiritual basis that can be a catalyst for personal change and transformation. Over the many years working with dogs, we have been struck repeatedly by how dogs mirror us back to ourselves in unmistakable ways. Because dogs are guileless and utterly themselves, they lack the capacity to deceive. If we take seriously the words they speak to us about ourselves, we stand face-to-face with our own truth. When we pay attention to these words inscribed on their bodies, in their expres-

*Saint Francis tames the wolf of Gubbio. Francis knew the secret of sensitive body language. Through his own deep awareness and appreciation of things, he unified the world of animals and human beings in himself.*

sions, in the way they approach and interact with us, they can stimulate a new level of consciousness.

For example, consider how a dog's eyes speak. They reflect a broad range of inner emotions that affect the quality of our relationship, if we care enough to listen. Joy, fear, curiosity, boredom, and mischief are each reflected through the subtlest shifts in the dog's eyes. It is crucial not to miss these cues. Just as the New Testament teaches that "the eye is the lamp of the soul" in human beings, so to a certain extent can the same be said of the dog.

This is not fantasy or New Age babble, but entirely consistent with scientific studies and the best examples of classic spirituality. Saint Francis of Assisi, for one, exemplifies this profound respect for the created world around us, particularly animals, and he has been a perennial source of inspiration to all, most especially to those who wish to intensify their communion with nature. Though his love for animals has often been romanticized to the point of sentimentality, beneath the legend we find a human being who was conscious of the mystery of the interconnectedness of everything and who expressed this in his attitude of reverence and wonder. Both people and animals responded to him in dramatic ways.

Perhaps Francis himself did not fully understand his skill with animals in any rigorous, scientific sense, but he was aware of his kinship with them. He did not leave us any "technique" to help us relate to animals more easily, but he obviously had the knack of communicating friendliness to animals through his own body language, something that we can easily recognize from our own scientific experience. He understood intuitively how to approach an animal so that the creature perceives the intruder as a friend rather than a threat. Francis's taming of the wild wolf of Gubbio resembles the way the prophet Daniel naturally communicated with the lions in their den, through the sensitive use of body language. These episodes point to a certain spiritual wholeness and integration as the keystones to relating optimally with animals, breaking down the traditional opposition between human and nonhuman creatures. By their focused inner attitude, these two saints unified the world of animals and human beings.

Francis saw the elements of nature, the cosmos, and all living creatures as his brothers and sisters. For this, if for no other reason, he is a model for dog owners and animal lovers, for the ecologist, for the naturalist, and for everyone who wishes to develop this kinship with life and respect for nature.

Although Francis was a Western Christian, his evocative personality and affinity with everything in creation is in perfect harmony with the theology of the Christian East, which does not arbitrarily divide the world of animal and human but strives to see the meaning and unity of creation. The annual Christian feast of the Transfiguration, central

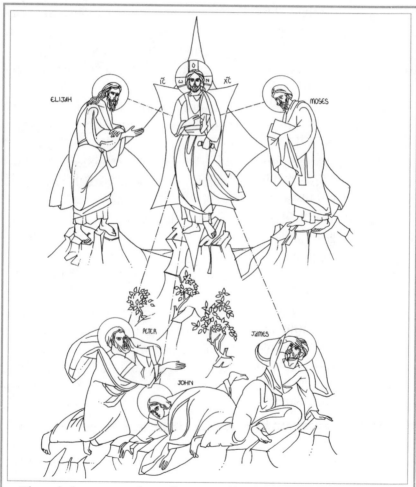

*Through the mystery of transfiguration (depicted here in the style of an ancient icon), all nature and humanity rises to a new level of awareness and harmony.*

in the liturgical tradition of the Eastern churches, clearly portrays this idea. When Jesus climbed to the top of Mount Tabor with three disciples, he was transfigured in a bright light there before their eyes. In fact, the whole mountain became beautiful and radiant with that light. Christian thinkers in the East have interpreted this as a kind of call to all creation, sentient and nonsentient alike, to rise to a new level of being. For us this means radically changing our thinking and behavior to meet the possibilities opened by this new reality in our individual lives and circumstances.

A Christian thinker of more modern times who stressed respect and compassion for animals is Father Pierre Teilhard de Chardin, the French Jesuit and paleontologist. Teilhard felt that "the mystical vibration is inseparable from the scientific vibration," something often underscored by many later physicists. He proposed that the entire universe is moving toward a cosmic unification and transformation, which he called the Omega point. Though his writings are not specifically concerned with dogs, they are of interest to the dog owner because they stress devotion and respect for the whole of creation.

As long as we arrogantly insist on separating the world of animals and the world of human beings, we will never be able to arrive at this heightened quality of consciousness that evokes from us genuine love, stewardship, and compassion. As this relates to our dogs, both dog and human being are the losers in such a scenario. But when we succeed in unifying the world of animals and human beings in ourselves through understanding, empathy, and training, then both have the chance to participate in a life that is more abundant and fulfilling.

This is a theme that appears again and again in monastic history. We have mentioned Saint Francis, but there are many others who developed an affinity with animals. Saint Antony and other spiritual fathers and mothers are said to have tamed lions and wolves. In Russia, Saint Sergius, Saint Seraphim, and others were on very friendly terms with the bears of the forest. Their monastic biographers interpreted this phenomenon as an illustration of the human task to help restore the order that existed in Paradise, the ideal natural order where human and beast, and indeed everything in creation, somehow live in harmony.

If we look to the book of Isaiah, we see the idea that human beings and animals can live together in peace is an age-old theme: "The wolf shall dwell with the lamb, and the leopard shall lie down with the kid, and the calf and the lion and the fatling together, and a little child shall lead them. . . . They shall not hurt or destroy in all my holy mountain; for the earth shall be full of the knowledge of the LORD as the waters cover the sea" (Isaiah 11:6, 7). Modern ideas of conservation, ecology, and animal training that emphasize careful stewardship and intelligent sensitivity wisely follow up this theme. As historian Kenneth Clark perceptively observed, "What is needed is not simply animal sanctuaries and extensive zoos, but a total change in our attitude. We must recognize that the faculty of speech which has given us power over those fellow creatures we once recognized as brothers must carry with it a proper measure of responsibility. We can never recapture the Golden Age, but we can regain that feeling of the unity of all creation. This is a faith we all may share."*

*If our relationship with our dog is to blossom to its fullest, our own sensitivity and awareness must be intensified. We can do so through our reading and by listening and seeing into the dog's own world. Only then will training, practice, praise, and play be most effective and rewarding.*

*Kenneth Clark, "Animals and Men: Love, Admiration, and Outright War," *Smithsonian* (September 1977), p. 57.

These modest essays in training presuppose that we love our dogs, but they also demand a great deal of thought and reflection from us. If our relationship with our dogs is to blossom to its fullest, we must cultivate our sensitivity and awareness. Often what prevents us from realizing the true potential in a relationship with a dog is our own lack of imagination of what can be. Good relationships demand creativity and patience, along with a self-discipline that is eager to draw the best out of the dog and ourselves. This is when training transcends the mundane to become spiritual and inspiring.

For such a vision to come to fulfillment, the invisible, ineffable current we call life must be the object of our love. Just as we ourselves share in it, so do other creatures, and herein lies great mystery. We now know that the responsibility for nurturing it falls to us. We humans alone can work out the delicate harmonies in this symphony, melodies composed in the key of life. If and when we do, we will indeed renew and enrich ourselves and the earth, even if we still fall short of fully regaining that golden age of perfect harmony.

# Select Reading List

The number of books on dogs, canids, and subjects relating to them are legion. Through the years we have benefited from the insights and experience of authors and trainers from many different perspectives. The following is a sampling of books we've found valuable.

GENERAL BOOKS

American Kennel Club. *The Complete Dog Book*. New York: Howell, 1998.
Bergler, Reinhold. *Man and Dog*. New York: Howell, 1988.
Bergman, Goran. *Why Does Your Dog Do That?* New York: Howell, 1971.
Boone, J. Allen. *Kinship with All Life*. New York: Harper and Row, 1954.
Budniansky, Stephen. *The Truth About Dogs*. New York: Viking, 2000.
Buytendijk, F. C. *The Mind of the Dog*. New York: Arno Press, 1973.
Caras, Roger. *A Dog Is Listening*. New York: Summit Books, 1992.
Donaldson, Jean. *The Culture Clash*. Oakland, Calif.: James and Kenneth Publishers, 1997.
Fox, Michael W. *Understanding Your Dog*. New York: Coward, McCann and Geoghegan, 1976.
Hearne, Vicki. *Adam's Task*. New York: Alfred A. Knopf, 1982.
Hoffman, Matthew, ed. *Dogs: The Ultimate Guide*. Emmaus, Penn.: Rodale Press, 1998.
Lorenz, Konrad. *Man Meets Dog*. Baltimore: Penguin Books, 1953.
Masson, Jeffrey Moussaieff. *Dogs Never Lie About Love*. New York: Three Rivers Press, 1997.
McSoley, Ray. *Dog Tales*. New York: Warner Books, 1988.
Morris, Desmond. *Dogwatching*. New York: Crown, 1986.
Pfaffenberger, Clarence. *The New Knowledge of Dog Behavior*. New York: Howell, 1968.
Sheldrake, Rupert. *Dogs That Know When Their Owners Are Coming Home*. New York: Three Rivers Press, 1999.
Thomas, Elizabeth Marshall. *The Hidden Life of Dogs*. Boston: Houghton Mifflin, 1993.
Trumler, Eberhard. *Your Dog and You*. New York: Seabury Press, 1973.

BOOKS ABOUT WOLVES AND OTHER WILD CANIDS

Bass, Rick. *The Ninemile Wolves*. New York: Ballantine Books, 1993.
Fox, Michael W. *The Behavior of Wolves, Dogs, and Related Canids*. New York: Harper and Row, 1970.

Lopez, Barry H. *Of Wolves and Men.* New York: Charles Scribner's Sons, 1978.

Mech, David. *The Arctic Wolf: Ten Years with the Pack.* New York: Voyageur Press, 1997.

———. *The Way of the Wolf.* New York: Voyageur Press, 1995.

———. *The Wolf: The Ecology and Behavior of an Endangered Species.* Minneapolis: University of Minnesota Press, 1981.

## BOOKS ON SELECTING A DOG

Caras, Roger. *The Roger Caras Dog Book.* New York: Holt, Rinehart and Winston, 1980.

Hart, Benjamin L. and Lynette A. *The Perfect Puppy: How to Choose Your Dog and Its Behavior.* New York: W. C. Freeman, 1988.

Kilcommons, Brian, and Sarah Wilson. *Paws to Consider.* New York: Warner Books, 1999.

Kilcommons, Brian, Sarah Wilson, and Michael Capuzzo. *Mutts: America's Dogs.* New York: Warner Books, 1996.

Siegal, Mordecai. *A Dog for the Kids.* Boston: Little, Brown, 1984.

Tortora, Daniel. *The Right Dog for You.* New York: Simon and Schuster, 1980.

## PUPPY CARE AND TRAINING

Monks of New Skete. *The Art of Raising a Puppy.* Boston: Little, Brown, 1991.

Pinkwater, Jill and Manus. *Superpuppy.* New York: Seabury Press, 1977.

Ross, John, and Barbara McKinney. *Puppy Preschool.* New York: St. Martin's Press, 1996.

Rutherford, Clarice, and David H. Neil. *How to Raise a Puppy You Can Live With.* Loveland, Colo.: Alpine Publications, 1981.

## OBEDIENCE TRAINING

Barwig, Susan, and Stewart Hilliard. *Schutzhund.* New York: Howell, 1991.

Bauman, Diane. *Beyond Basic Dog Training.* New York: Howell, 1987.

Benjamin, Carol Lea. *Mother Knows Best.* New York: Howell, 1987.

Booth, Sheila. *Purely Positive Training.* Ridgefield, Conn.: Podium Publications, 1998.

Burnham, Patricia. *Playtraining Your Dog.* New York: St. Martin's Press, 1980.

Frankel, Cis. *Urban Dog.* Minocqua, Wisc.: Willow Creek Press, 2000.

Frost, April. *Beyond Obedience.* New York: Three Rivers Press, 1998.

Kilcommons, Brian, with Sarah Wilson. *Good Owners, Great Dogs.* New York: Warner Books, 1992.

Koehler, W. R. *The Koehler Method of Dog Training.* New York: Howell, 1969.

Most, Konrad. *Training Dogs.* London: Popular Dogs, 1974.

Pryor, Karen. *Don't Shoot the Dog.* New York: Bantam Books, 1985.

Ross, John, and Barbara McKinney. *Dog Talk*. New York: St. Martin's Press, 1992.

Spector, Morgan. *Clicker Training for Obedience*. Waltham, Mass.: Sunshine Books, 1999.

Strickland, Winifred. *Expert Obedience Training for Dogs*. New York: MacMillan, 1987.

Tucker, Michael. *Dog Training Made Easy*. Adelaide: Rigby Publishers, 1980.

Volhard, Joachim, and Gail Tarmases Fisher. *Training Your Dog*. New York: Howell, 1983.

Volhard, Joachim and Wendy. *The Canine Good Citizen*. New York: Howell, 1997.

———. *The Complete Idiot's Guide to a Well-Trained Dog*. New York: Alpha Books, 1999.

Woodhouse, Barbara. *No Bad Dogs*. New York: Summit Books, 1982.

## BEHAVIOR PROBLEMS

Benjamin, Carol Lea. *Dog Problems*. New York: Doubleday, 1981.

Campbell, William E. *Behavior Problems in Dogs*. 3d rev. Grants Pass, Ore.: BehavioRx Systems, 1999.

Carlson, Jeanne, and Ranny Green. *Good Dogs, Bad Habits*. New York: Fireside, 1995.

Dodman, Nicholas. *The Dog Who Loved Too Much*. New York: Bantam, 1996.

———. *Dogs Behaving Badly*. New York: Bantam, 2000.

Siegal, Mordecai, and Matthew Margolis. *When Good Dogs Do Bad Things*. Boston: Little, Brown, 1986.

Tortora, Daniel. *Help! This Animal Is Driving Me Crazy*. New York: Wideview, 1978.

## BOOKS OF A MORE TECHNICAL OR SPECIALIZED NATURE THAT YOU MAY BE INTERESTED IN

Burch, Mary, and Jon Bailey. *How Dogs Learn*. New York: Howell, 1999.

Coppinger, Raymond and Lorna. *Dogs: A Startling New Understanding of Canine Origin, Evolution and Behavior*. New York: Simon and Schuster, 2001.

Fiennes, Richard and Alice. *The Natural History of Dogs*. New York: Bonanza, 1968.

Fogle, Bruce. *The Dog's Mind*. New York: Howell, 1992.

Fox, Michael. *Integrative Development of Brain and Behavior in the Dog*. Chicago: University of Chicago Press, 1971.

Milani, Myrna. *The Body Language and Emotions of Dogs*. New York: Quill, 1986.

Scott, Jonn Paul, and John L. Fuller. *Genetics and the Social Behavior of the Dog*. Chicago: University of Chicago Press, 1965.

Serpell, James, ed. *The Domestic Dog*. Cambridge: Cambridge University Press, 1995.

Volhard, Wendy, and Kerry Brown. *The Holistic Guide for a Healthy Dog.* New York: Howell, 1995.

### VIDEOTAPES

*Raising Your Dog with the Monks of New Skete* (set of three tapes). Available from the monastery through www.newskete.com.

### MAGAZINES

AKC Gazette
American Kennel Club
260 Madison Avenue
New York, NY 10016
www.akc.org

Dog World Magazine
500 N. Dearborn, Suite 1100
Chicago, IL 60610
www.dogworldmag.com

Front & Finish
The Dog Trainer's News
P.O. Box 333
Galesburg, IL 61402-0333
www.frontfinish.com

Off-Lead Magazine
The Dog Training Instructors Magazine
Barkleigh Productions, Inc.
6 State Road #113
Mechanicsburg, PA 17050

### WHERE TO GET TRAINING EQUIPMENT

Handcraft Collars, Inc.
4875 Camp Creek Road
Pell City, AL 35125
(800) 837-2033
www.handcraftcollars.com
This is where we obtain the Volhard collars described in the chapter on equipment.

J-B Wholesale
5 Raritan Road
Oakland, NJ 07436
(800) 526-0388

# Appendix

## AKC Titles and Abbreviations*

AS A PREFIX

Conformation
    Ch.: Champion
Obedience
    NOC: National Obedience Champion
    OTCH: Obedience Trial Champion
Tracking
    CT: Champion Tracker (TD, TDX, and VST)
Agility
    MACH: Master Agility Champion
    MACH2, MACH3, MACH4, etc. MACH may be followed by a
    number designation to indicate the quantity of times the dog
    has met the requirements of the MACH title.
Field Trials
    FC: Field Champion
    AFC: Amateur Field Champion
    NFC: National Field Champion
    NAFC: National Amateur Field Champion
    NOGDC: National Open Gun Dog Champion
    AKC GDSC: AKC Gun Dog Stake Champion
    AKC RGDSC: AKC Retrieving Gun Dog Stake Champion
Herding
    HC: Herding Champion

*Taken from www.akc.org, the website of the American Kennel Club.

Dual

　　　DC: Dual Champion (Ch. and FC)

Triple

　　　TC: Triple Champion (Ch., FC, and OTCH)

Coonhounds

　　　NCH: Nite Champion

　　　GNCH: Grand Nite Champion

　　　SGNCH: Senior Grand Nite Champion

　　　GCH: Grand Champion

　　　SGCH: Senior Grand Champion

　　　GFC: Grand Field Champion

　　　SGFC: Senior Grand Field Champion

　　　WCH: Water Race Champion

　　　GWCH: Grand Water Race Champion

　　　SGWCH: Senior Grand Water Race Champion

AS A SUFFIX

Obedience

　　　CD: Companion Dog

　　　CDX: Companion Dog Excellent

　　　UD: Utility Dog

　　　UDX: Utility Dog Excellent

　　　VCD1: Versatile Companion Dog 1

　　　VCD2: Versatile Companion Dog 2

　　　VCD3: Versatile Companion Dog 3

　　　VCD4: Versatile Companion Dog 4

　　　VCCH: Versatile Companion Champion

Lure Coursing

　　　JC: Junior Courser

　　　SC: Senior Courser

　　　MC: Master Courser

Tracking

　　　TD: Tracking Dog

　　　TDX: Tracking Dog Excellent

　　　VST: Variable Surface Tracker

Agility

　　　NA: Novice Agility

　　　OA: Open Agility

AX: Agility Excellent

MX: Master Agility Excellent

NAJ: Novice Jumpers with Weaves

OAJ: Open Jumpers with Weaves

AXJ: Excellent Jumpers with Weaves

MXJ: Master Excellent Jumpers with Weaves

Hunting Test

JH: Junior Hunter

SH: Senior Hunter

MH: Master Hunter

Herding Test

HT: Herding Tested

PT: Pre-Trial Tested

HS: Herding Started

HI: Herding Intermediate

HX: Herding Excellent

Lure Coursing

JC: Junior Courser

SC: Senior Courser

MC: Master Courser

Earthdog

JE: Junior Earthdog

SE: Senior Earthdog

ME: Master Earthdog

INTERNATIONALLY RECOGNIZED TITLES AWARDED BY USA JUDGES*

SchHA: Introduction to Schutzhund work without the tracking

Bh: Basic companion dog

WH: Watch dog test for basic alertness

AD: Endurance test for fundamental fitness

SchH 1: The preliminary Schutzhund qualification in tracking, obedience, and protection

SchH 2: More challenging Schutzhund work in tracking, obedience,
and protection

*Taken from www.germanshepherddog.com, a website with loads of informa-
tion about Schutzhund and other working degrees.

SchH 3: The competition level of the three phases of Schutzhund. Masters level.

FH 1: Advanced tracking

FH 2: Greater tracking challenges. Placement of articles determined by judge.

IPO 1: International trial rules similar to the Schutzhund test, but with some variations.

IPO 2: More challenging Schutzhund work in tracking, obedience, and protection.

IPO 3: The competition level of IPO.

ADDITIONAL SV TITLES RECOGNIZED BY USA
(U.S. SCHUTZHUND CLUB)

BpDH 1 & 2: Railway Police Dog

BIH: Blind Leader Dog

DH: Service Dog

DPH: Service Police Dog

HGH: Herding Dog

IPO 1, 2, 3: International Rules (same as USA)

LwH: Avalanche dog

PFP 1 & 2: Police Tracking Dog

PH: Police Dog

PSP 1, 2, 3: Police Guard Dog

RtH: Rescue Dog

SchH 1, 2, 3: Schutzhund titles (same as USA)

ZFH: Customs Tracking Dog

ZH 1, 2, & 3: Customs Dog

# Index